Sustainable Agricultural Practices

Sustainable Agricultural Practices: Economic and Environmental Implications

EDITED BY

RAMESH CHANDRA DAS

Vidyasagar University, India

emerald PUBLISHING

United Kingdom – North America – Japan – India – Malaysia – China

Emerald Publishing Limited
Emerald Publishing, Floor 5, Northspring, 21-23 Wellington Street, Leeds LS1 4DL.

First edition 2025

British Library Cataloguing in Publication Data
A catalogue record for this book is available from the British Library

ISBN: 978-1-83608-337-5 (Print)
ISBN: 978-1-83608-336-8 (Online)
ISBN: 978-1-83608-338-2 (Epub)

Printed and bound by CPI Group (UK) Ltd, Croydon, CR0 4YY

INVESTOR IN PEOPLE

In memory of my teacher
Professor Rabindranath Bhattacharya

Contents

List of Abbreviations

2SLS:	Two-stage least squares
ABP:	Anchor Borrowers' Programme
ADF:	Augmented Dickey–Fuller
AEC:	Architecture, engineering and construction
AGMARKNET:	Agricultural Marketing Information Network
AGS:	Agricultural production by sector
APMC:	Agricultural Produce & Livestock Market Committee
AR:	Auto-regression
ARCH:	Auto-regressive conditional heteroskedasticity
ARDL:	Auto-regressive distributive lag
ASI:	Agricultural Sustainability Index
BIMSTEC:	Bay of Bengal Initiative for Multi-Sectoral Technical and Economic Cooperation
CBN:	Central Bank of Nigeria
CII:	Coppock's Instability Index
CO_2:	Carbon dioxide
CO_2 emission:	Carbon dioxide emission
CPI:	Consumer price index
CR:	Convergence rate
CRP:	Crop production
CV:	Coefficient of variation
CVM:	Contingent valuation method
DAP:	Di-ammonium phosphate
DBIE:	Database on Indian Economy
DC:	Dichotomous choice
DEA:	Data envelopment analysis
DEMATEL:	The decision-making trial and evaluation laboratory
DFS:	Dhaka Food System
DOLS:	Dynamic ordinary least square
DW:	Durbin Watson
ECSI:	Economic Sustainability Index
ECT:	Error correction term
EG:	Economic globalization
EKC:	Environmental Kuznets curve
ENSUB:	Environmental sustainability
ESI:	Environmental Sustainability Index

FAO:	Food and Agriculture Organization of the United Nations
FDI:	Foreign direct investment
FFW:	Food for work
FMOLS:	Fully modified ordinary least square
FW:	Food waste
FY:	Fiscal year
GAIN:	Global Alliance for Improved Nutrition
GARCH:	Generalized auto-regressive conditional heteroskedasticity
GDP:	Gross domestic product
GFCF:	Gross fixed capital formation
GHGs:	Greenhouse gases
GIB:	Green innovation business
GJR:	Glosten–Jagannathan–Runkle
GMA:	Gross margin analysis
GMM:	Generalized method of moments
GVA:	Gross value added
GXA:	Total government expenditure
HL:	Half-life of convergence
IFPRI:	International Food Policy Research Institute
IMR:	Infant mortality rate
IPCC:	Intergovernmental Panel on Climate Change
IPM:	Integrated pest management
IPS:	Im, Pesaran, and Shin
IRF:	Impulse response function
LAB:	Population growth
LLC:	Levin, Lin, and Chu
LRI:	Live Rice Index
LSP:	Livestock production
MC:	Marginal cost
MLE:	Maximum likelihood estimates
MSC:	Marginal social cost
M-TAR:	Momentum-threshold auto-regressive
NABARD:	National Bank for Agriculture and Rural Development
NCA:	Necessary condition analysis
NCRMP:	National Cyclone Risk Mitigation Project
NGO:	Non-governmental organization
NTFPs:	Non-timber forest products
OECD:	Organization for Economic Co-operation and Development
OLS:	Ordinary least square
OMS:	Open market sale
PCA:	Principal component analysis
PCAGDP:	Per capita agricultural GDP
PCFW:	Per capita food waste
PCGDP:	Per capita GDP
PGDP:	GDP per capita
POCAA:	Platform of Community Action and Architecture

PRSSP:	Policy Research and Strategy Support Programme
RBI:	Reserve Bank of India
RCA:	Revealed comparative advantage
RRB:	Regional Rural Bank
RUM:	United Nations Educational, Scientific and Cultural Organization
SCBs:	Schedule commercial banks
SDG:	Sustainable development goals
SFR:	Stochastic frontier regression
SLR:	Sea-level rise
SPS:	Sanitary and phytosanitary standard
SSI:	Social Sustainability Index
SWOT:	Strengths, weaknesses, opportunities and threats
TAP:	Total agricultural production
TE:	Technical efficiency
UA:	Urban agriculture
UN SDGs:	United Nations Sustainability Development Goals
UN:	United Nations
UNDP:	United Nations Development Programme
VAR:	Vector auto regression
VECM:	Vector error correction mechanism
VGF:	Vulnerable group feeding
VIF:	Variance inflationary factor
WDI:	World Bank Development Indicator
WFP:	World Food Programme
WTO:	World Trade Organisation
WTP:	Willingness to pay

About the Editor

Ramesh Chandra Das, PhD, is presently a Professor and Head of the Department of Economics, Vidyasagar University, India, with 25 years of teaching and research experience. He has published more than 100 articles in reputed journals and edited books and more than 20 books on his credit. His research interests lie in applied macroeconomics, public economics and environmental economics.

About the Contributors

Ugochukwu D. Abasilim, PhD, is a Senior Lecturer of Public Administration at the Department of Political Science and international Relations, Covenant University, Ota, Ogun State, Nigeria, with over a decade of experience. His expertise lies in educating students and conducting research in public administration, public personnel management, public service ethics and e-government.

Ebikabowei Biedomo Aduku, PhD, is now a Lecturer at the Department of Economics, University of Africa, Nigeria. Earlier, he was a Researcher at the Department of Economics, University of Nigeria, Nsukka, Nigeria. He has attended several national and international conferences. He has publications in books and journals of reputation.

Kishan Agarwalla is currently working as a Senior Research Fellow at the Department of Economics, University of North Bengal. He is also attached to the Department of Economics at the Indian Institute of Legal Studies, Siliguri, India, as Faculty. His areas of research interest includes resource economics, environmental economics, mathematical economics, statistics and international economics.

Md. Saifullah Akon is an Assistant Professor in Japanese Studies at the University of Dhaka. He obtained his Bachelor's and Master's degrees in International Relations at the University of Dhaka. He has successfully participated in a short course winter programme at Sophia University, Tokyo, Japan in January 2019. His areas of interest include foreign policy, security, Japan–South Asia relations and disaster governance.

Mainak Bhattacharjee is presently an Assistant Professor in Economics at Loreto College, Kolkata, India. He has also worked in the same post in The Heritage College, Kolkata, India. He has obtained MPhil and MA in Economics, respectively, from the Jadavpur University. He has been working in the areas of macroeconomics and international trade. He has produced many research articles published in reputed journals.

Kiran Bhowmik is a Research Scholar at the Department of Economics pursuing PhD in Agricultural Economics. Earlier, he completed his post-graduation in Economics from Tripura University (A Central University).

Çağatay Çağlayan is currently pursuing the Master's degree in Business Management at Istanbul Medipol University. His research interests include sustainable energy economics, renewable energy and nuclear energy.

Nilendu Chatterjee, PhD, is an Assistant Professor in the Department of Economics, Bankim Sardar College, West Bengal, India. He has research interest in resource economics, general equilibrium and development economics. He has published a number of research articles in several international journals of economics including *International Journal of Sustainable Economies Management*, *Economic Affairs* and *Foreign Trade Review*.

Tonmoy Chatterjee, PhD, is an Assistant Professor in the Department of Economics, Bhairab Ganguly College, West Bengal, India. He has research interest in International Trade Theory, General Equilibrium, Health Economics and Development Economics. He has published a number of research articles in several international journals of economics.

Obinna S. Chima holds PhD in Policy and Strategic Studies from the Department of Political Science and International Relations, Covenant University, Ota, Ogun State, Nigeria. He is also the Editor of *THISDAY* Newspaper, has authored extensive reports and has actively participated in meetings of the International Monetary Fund and World Bank. His research interests are food security, public policy and governance and economic and strategic studies.

Debraj Das is a Doctoral-level Research Scholar at the Department of Economics, West Bengal State University and Assistant Professor in Economics at Kabi Jagadram Roy Government General Degree College, West Bengal. His research interests lie in agricultural economics, social sector and development economics.

Subhrabaran Das, PhD, is presently serving at Tripura University (Central University) as Professor and Head, Department of Economics. Earlier, he served as an Assistant Professor at Assam University (Central University), Silchar and engaged in post-graduate teaching at Presidency College, Kolkata. He has done research work on different issues such as social sector issues, gender issues, industry and agricultural issues. He has published more than 30 research papers in different reputed journals and has 3 books in his credit.

Buddhadev De is a Doctoral-level Research Scholar at the Department of Economics, Vidyasagar University, India. His research interests lie in agricultural economics, economics of food waste and poverty, development economics, etc.

Moses M. Duruji, PhD, is a Professor of Political Science with 20 years of experience at the Department of Political Science and International Relations, Covenant University, Ota, Ogun State, Nigeria. His areas of specialisation include social movements, nationalism studies, ethnic nationalism, democracy studies and education.

Ujjal Protim Dutta, PhD, currently serves as an Assistant Professor in the Department of Humanities and Social Sciences at the National Institute of Technology Silchar, Assam, India. Previously, he held the position of Assistant Professor in the Department of Economics at Rangia College, Assam. He earned his PhD from the National Institute of Technology Durgapur and holds a Master's degree in Development Studies from the Indian Institute of Technology Guwahati.

İrfan Ersin, PhD, is an Assistant Professor at Istanbul Medipol University, Turkey. He finished his Doctoral programme at Medeniyet University in 2022. His research interests are development economics, macroeconomics, inflation and employment.

Serkan Eti, PhD, is currently an Assistant Professor with İstanbul Medipol University, Turkey. He received PhD degree from Marmara University in 2016. His research interests include data mining, fuzzy logic and multi-criteria decision-making methods.

Daniel E. Gberevbie, PhD, is a Professor of Public Administration with 21 years of experience at the Department of Political Science and International Relations, Covenant University, Ota, Ogun State, Nigeria. His areas of specialisation are public administration, ethical issues, employee retention, e-governance and gender politics.

Subhabrata Ghosh, PhD, has been working as an Assistant Professor in the Department of Botany, Mahadevananda Mahavidyalaya, Barrackpore, India for over eight years. He is actively involved in research in the area of Algal morpho-taxonomy and ecology. He has published over 20 research papers in reputed peer-reviewed journals.

Yaşar Gökalp, PhD, is an Assistant Professor at Istanbul Medipol University, Turkey. He finished his Doctoral programme at Istanbul Medipol University in 2022. His research interests are energy, health economics, health finance and multi-criteria decision-making techniques.

Md. Shihabul Hossain, a Senior Officer at BRAC, earned his Bachelor of Social Science in Development Studies from Khulna University and a Master's in Disaster Management from the University of Dhaka. Leveraging his academic and professional background, he is devoted to working as a development professional in the environmental and humanitarian arena at national and international organizations.

Aliul Islam is a Research Scholar specializing in Economics at the National Institute of Technology Silchar, Assam, India. He holds a Master's degree in Economics from Guwahati University, Guwahati, Assam. His research interests primarily focus on financial markets, delving into how various economic forces shape and impact market dynamics.

Enrico Ivaldi is Associate Professor in Social Statistics, in the Faculty of Communication, at the Free University of Languages and Communication IULM, Milan branch. He holds a PhD in Applied Economics and Quantitative Methodologies from the University of Genoa. He is a member of the Editorial Board of *International Journal of Transport Economics, Revista de Estudios Andaluces, Risk and Vulnerability*, curriculum in Management and Security, University of Genoa. He is also a member of the Centro de Investigaciones en Econometría of the Universidad de Buenos Aires, Argentina, and of the Pontificia Accademia Mariana Internazionale, Department of Integral Welfare, Vatican City. He is a member of several scientific societies, including the Italian Statistical Society and the Royal Statistical Society. His main areas of scientific interest are the analysis of social indicators, the study of deprivation and inequality, and smart and sustainable cities.

Sagnik Maity is a Doctoral Research Scholar (UGC-JRF) in the Department of Commerce, University of Calcutta. His research mainly revolves around the area of finance. He holds a keen interest in financial inclusion and financial technology. He has presented research papers in several national and international conferences and published articles and papers in various reputed journals.

Amit Majumder, PhD, is presently working as Associate Professor of Commerce and PG Coordinator of Bijoy Krishna Girls' College, India. He is also a Visiting Faculty, Department of Commerce, University of Calcutta. He has authored three books from international publisher and contributed in 60 edited volumes, 50 nationally and internationally acclaimed peer-reviewed and referred journals. His areas of interest are strategic management, corporate governance, etc.

Debashis Mazumdar, PhD, is working as Professor of Economics at The Heritage College, Kolkata. He is also acting as Visiting Faculty in the Post-Graduate Department of Commerce at Nababullygunge Mahavidyalaya Kolkata and in the PG (Economics) Section of Indira Gandhi National Open University, Muralidhar Girls' College Center, Kolkata. He has written a good number of text books on managerial economics and contributed articles in many national and international journals.

Md. Juel Mia is an Assistant Professor at the Institute of Disaster Management and Vulnerability Studies, University of Dhaka, Bangladesh. Prior to his career at the university, he served as a newsroom editor at Somoy News (a television channel in Bangladesh). He is the Founder and Chairperson of Disaster Perception, a volunteer platform. His research interests are disaster communication, environmental geography, geospatial analysis and disaster governance.

Debabrata Mukhopadhyay, PhD, is a Professor in the Department of Economics at the West Bengal State University, India. He has published a number of papers in scholarly national and international journals. He has obtained his PhD degree in Quantitative Economics from the Indian Statistical Institute in 2008.

Shamima Prodhan is an Assistant Professor at the Institute of Disaster Management and Vulnerability Studies, University of Dhaka, Bangladesh. Her research interests include urban disaster and management, water security, nature-based solutions and climate change adaptation. She has also authored several journal articles related to water security, river pollution, water intervention, COVID-19, salinity, adaptation strategies and many more.

Begum Sertyesilisik is Professor in the Faculty of Architecture at the Istanbul University, Türkiye. She has received her PhD degree from the Middle East Technical University, and her MSc, MBA and BSC degrees from the Istanbul Technical University. Her specialization areas include green marketing, sustainability, sustainable development and sustainable built environment.

Krishna Singh serves as an Assistant Professor of Economics at the University of Gour Banga, with a specialization in Econometrics and Statistics. He holds a PhD in Economics from Burdwan University. He has published more than 20 papers in various national and international journals of repute.

M. C. Omar C. Vargas-González is Professor and Head of the Department of Systems and Computing at the Technological National of Mexico Campus Ciudad Guzmán, México. His areas of research interest lie in business administration, entrepreneurial economics, computing in social science, arts and humanities and mathematics education.

José G. Vargas-Hernández, PhD, is presently a Research Professor, Tecnològico Nacional de Mèxico /ITSF. He has a long career in Administrative and Management Sciences having more than 200 articles published in different journals, books and reports.

Duygu Yavuz is a Master's student at Istanbul University. She completed her Bachelor's degree in Health Management (English) and double major in Economics and Finance (English). Her research areas are energy, finance, energy economy and sustainability.

Foreword

In the face of growing global challenges such as climate change, population growth and food insecurity, sustainable agricultural practices emerge not only as a choice but also as an imperative. The present book titled *Sustainable Agricultural Practices: Economic and Environmental Implications* stands as a beacon in the international landscape, offering profound insights into the intersection of economics, environment and agriculture. Moreover, the interdisciplinary nature of the contributions underscores the interconnectedness of economic, environmental and social dimensions in agriculture. By integrating perspectives from fields such as economics, environmental science and agriculture, this book fosters a holistic understanding of the challenges and opportunities facing agricultural sustainability.

In this meticulously curated anthology, scholars from diverse backgrounds and regions converge to dissect the many dimensions of sustainable agriculture. Divided into two sections, the book navigates through the economic and environmental implications of agricultural practices, providing a comprehensive understanding of the contemporary agricultural landscape.

The first section, 'Economic Implications of Agricultural Practices in Countries and Groups', delves into the economic complexities that characterize agricultural systems around the world. From examining productivity changes in cereal crops in BIMSTEC countries to examining the role of institutional credit in the development of India's agricultural sector, each chapter offers a nuanced perspective on the economic forces driving agricultural dynamics. In addition, the section illustrates how modern agricultural techniques and entrepreneurial models can catalyse agricultural transformation, as shown in studies from Mexico to Nigeria.

In the second section, 'Environmental Insinuations of Agricultural Practices Across Countries and Groups', the focus shifts to the environmental impacts of agricultural activities. Through empirical investigations and theoretical frameworks, contributions clarify strategies to mitigate greenhouse gas emissions, address environmental degradation and promote sustainable agricultural exercises. From Turkey to Bangladesh, chapters emphasize the global relevance of sustainable agricultural practices to safeguard environmental health and ensure food security.

The breadth and depth of topics covered in this volume underscore the interdisciplinary nature of sustainable agriculture. Scholars from economics, environmental science, agriculture and related fields converge to provide a holistic understanding of the challenges and opportunities facing modern agriculture.

Moreover, by highlighting case studies from diverse geographic contexts, the book underscores the universality of the imperatives of sustainable agriculture and the need for context-specific solutions.

Finally, this book is not only an excellent scientific compendium but also a call to action. It invites policymakers, practitioners, researchers and students to engage in dialogue, innovation and collaboration to forge a more sustainable agricultural future. By integrating economic viability with environmental protection, we can achieve agricultural systems that nourish both people and the planet.

Enrico Ivaldi
IULM University, Milan, Italy
enrico.ivaldi@iulm.it

Preface

The need for food has increased in tandem with the growth in the world's population. In addition to causing environmental harm, the agricultural sectors of the majority of nations have now adopted modern technologies that yield high yield rates. Thus, modern farming methods have both positive and negative effects on the economy and the environment. A number of contaminants combined with agricultural output cast doubt on the sustainability issue's effectiveness. Large-scale agricultural production can release additional greenhouse gases and more hazardous pollutants. Once again, these emissions may have an impact on weather variability, increase the likelihood of climate change and increase the rate of global warming. This may have a greater detrimental impact on the dairy cow feeding industry than on other agriculture-related businesses. Consequently, when the likelihood of a disaster increases, so too may the vulnerability of livelihoods depending on agriculture. But there isn't a reliable and solid model that can calculate the social costs associated with these emissions. Therefore, reducing the dangers of pollution and environmental degradation as well as adapting to climate change and weather unpredictability should not only make farming systems more resilient but also stabilize farmers' output and revenue.

Under the ambiance, the present edited book titled *Sustainable Agricultural Practices: Economic and Environmental Implications* pitches light upon accumulating studies related to economic and environmental aspects of agricultural practices over the years across countries and groups of the world.

The book has compiled of 16 chapters which are divided into two broad sections. Section A covers up the Economic Implications of Agricultural Practices in Countries and Groups containing nine chapters including this introductory chapter and Section B covers up the studies related to the Environmental Insinuations of Agricultural Practices Across Countries and Groups casing seven remaining chapters.

The summary observations are that the countries have experienced rising growth of agricultural sector leading to the improvements in the overall economic activities. There are so many institutional supports from banks, several borrowing programmes and entrepreneurship developments, among others, which have worked as the catalyst for the additional growth progress in the agricultural sector. It has also been observed that the developed countries with high per capita income and relatively low dependence on agriculture sector do waste more food due to their preferences over good quality foods compared to the so-called low income countries. Besides the magnificent achievements in growth front, the activities in the sector have led to the environmental degradation although a list of studies has focussed on recommending the climate change issues. Zero budgeting towards inorganic fertilizers usage and full budgeting towards organic farming practices is also recommended. Thus, as the concluding observation, the entire book has revealed the spectacular growth of the agriculture sector during the modern times with some sort of negative externalities in terms of environmental degradation. Both developed and less developed economies should focus upon sustainable agricultural practices, as has been recommended by the listed research on the topic to have a better future so that inter-generational equities can be maintained.

The contents of the book may provide thought-provoking solutions to the existing problems – economic and environmental aspects of agricultural practices in order to achieve sustainable development in true sense. The readers and policy-makers will be immensely benefitted from the contents of the book.

While executing the book project, the editor got marvellous support from different angles of the academic and social world. The editor acknowledges the support of the entire Emerald Team, the contributing authors and the Foreword writer. Moreover, the editor is beholden to his parents, wife, daughter and other members of the family for their sacrifice and cooperation to divulge the volume. Though, the editor is exclusively accountable for any error that still remains in the book.

Ramesh Chandra Das

Section A

Economic Implications of Agricultural Practices in Countries and Groups

Chapter 1

Economic and Environmental Implications to Sustainable Agricultural Practices: Insights from the Covered Studies

Ramesh Chandra Das

Department of Economics, Vidyasagar University, India

Abstract

With the rise in the global population size there has been a parallel increase in the food demand. The agricultural sector in most of the countries has now come up with the uses in the modern technologies resulting into high yield rates besides environmental damage. The new agricultural practices of today have thus both economic positivity and environmental negativity. The present book intends to investigate the growth of agricultural sector and its implications to national output and environmental sustainability. It covers up economic and environmental implications of the agricultural practices in the modern world across the nations and groups of nations. The summary output of the coverage of the entire book shows that modern agricultural practices have led to high growth of the agricultural sector but environment has lost its in situ feature leading to sustainability problems. The recommendations are offered in terms of environmentally sustainable agricultural practices which can mitigate the challenges of climate change.

Keywords: Agricultural practices; productivity; climate change; food waste; zero budgeting; entrepreneurship; food waste; sustainability

Introduction

The world now is with a population load of around 800 crore and a giant share of this belongs to the Asia and African continents. With the rise in the population size there has been a parallel increase in the food demand. The agricultural

Sustainable Agricultural Practices: Economic and Environmental Implications, 3–9

doi:10.1108/978-1-83608-336-820241001

sector in most of the countries has now come up with the uses in the modern technologies with chemical fertilizers, pesticides and insecticides. There has been application of crop insurance against agricultural credit which leads to secured production leading to increase in supply chains involving exports and imports of agricultural produces. The countries' gross domestic product (GDP) in the head of agriculture sector is going up over the years, although the share of the agricultural sector in total GDP in most of the countries are going down (Čolović & Petrović, 2014; Das, 2023; Das et al., 2018; Duxbury, 2001; Enjolras et al., 2012; Gangopadhyay, 2004).

Amalgamation of several pollutants, particularly the methane emission, with agriculture production creates doubt over the efficacy of sustainability issue. Massive agriculture production can emit more vulnerable pollutants along with other greenhouse gases. Again, such methane emissions may affect weather variability and multiply climate change risks and the magnitude of global warming. This can affect dairy cattle feeding sector along with other agriculture-based activities more severely. As a consequence, the vulnerability of agriculture-based livelihoods may increase with induced disaster risks. However, there is no definite and robust model which can estimate social costs from such emissions. Hence, by minimizing environmental degradation and pollution risks along with adaptation to climate and weather variability risks should not only increase resilience of farmers' production systems but also stabilize their output and income. Identification and reduction of above-stated uncertainties and risk factors in terms of anticipatory adaptation may raise the potentiality of sustainability paradigm (Das & Mukherjee, 2020; Haddad, 2005; Hope, 2008; Volenzo, 2019). Therefore, climate change adaptation policies in the agricultural sector along with adaptation to control methane emissions are to be implemented for getting sustainable development.

In some parts of the world, the agricultural transformation from the low-productive natural sources to high-productive technology-based sources is known as the 'green revolution' which has made a vast part of the world, particularly of the Asia and African zones, food-secured. It is not that the cent per cent population is now out of the poverty network. Though a vast part of the world is food-secured, the technology-based modern agricultural practices have contributed a huge number of pollutants (such as carbon emission, methane, nitrogen, etc.) to the ambient environment making the growth and development process unsustainable.

Objectives of the Book

It is in this juncture, this book titled *Sustainable Agricultural Practices: Economic and Environmental Implications* aims to emphasize upon the economic and environmental aspects of agricultural practices over the years across countries and groups of the world. The readers from the disciplines such as Pure Economics, Agricultural Economics and Agricultural Science and Environmental Science are expected to be the main beneficiaries of the contents of the book.

Outcomes of the Chapters and Their Analysis

The book has the assembling of 16 chapters that are divided into two broad sections. Section I covers up the Economic Implications of Agricultural Practices in Countries and Groups containing nine chapters including this introductory chapter, and Section II covers up the studies related to the Environmental Insinuations of Agricultural Practices across Countries and Groups casing seven remaining chapters. The brief sketch of all the chapters is outlined below.

Chapter 2 attempts to analyse the change in productivity and cropping pattern among the Bay of Bengal Initiative for Multi-Sectoral Technical and Economic Cooperation (BIMSTEC) countries that comprises with Bangladesh, Bhutan, India, Myanmar, Nepal, Sri Lanka and Thailand. The study also examines the performance of BIMSTEC countries in exporting the agricultural products. For the empirical investigations the study uses co-efficient of variation, Coppock's instability index and Balassa's index. The study reveals that productivity and cropping pattern of the cereal production have been changing gradually over time in all BIMSTEC countries which shows the importance of products substitution. It also reveals that the development cooperation among the BIMSTEC countries through promotion of exports and imports is required to meet their needs.

Chapter 3 has attempted to relate the socio-economic aspects to the valuation of two most used as well as popular natural resources on which people of Sundarbans largely depend for survival – agriculture and fishery applying the very popular and appropriate methodology for valuation of natural resources – contingent valuation method – to get an idea about the valuation of these two resources by the stakeholders of the region. It is observed that people of Sundarbans are not only steadily reliant upon the use of these two resources for livelihood generation but they are very much eager to protect and sustainably use these resources as well, obviously except the problem of 'illegal bottom trawling'.

Chapter 4 investigates the fluctuations and instability of the two crops, paddy and wheat, in the states of India for the period 2006–2023 using coefficient of variation (CV), Cuddy-Della Valle index and Coppock's index. Then it estimates volatility by GARCH framework. It is observed that the average standard deviation of prices across mandis for paddy in a given month is 0.18 and for wheat it is 0.22. Both for paddy and wheat Punjab has the maximum instability among the three states if we consider CV and Coppock's index. But if we consider the Cuddy-Della Valle index, Madhya Pradesh has the maximum instability for wheat. The observed volatility in the price series of paddy revealed that the value of ARCH term has significant term and value of GARCH term is also significant for all the price series.

Chapter 5 investigates the evolving dynamics of India's agricultural sector, particularly the shift from reliance on informal money lenders to a more prominent role of institutional credit in recent decades employing the Vector Autoregressive methodology on data spanning from 1990 to 2022. To address the issue, researchers used direct institutional credit and the share of gross value-added (GVA) of agriculture and allied sectors at current prices, finding that neither of them causes the other. Further analysis concludes that in the future, the contribution of credit will have some impact on development.

Chapter 6 aims to examine the market of fertilizers in México during the pandemic crisis using the analytical-descriptive and the critical reflexive based on the quantitative data which points out that the consumption of fertilizers has undergone a change in the structure in favour of consumers with the highest concentration and diversification. To take advantage of the opportunities arising from the crisis, it analysed the functioning of an integral model of entrepreneurship in green innovation business (GIB) that is currently emerging and in the process of internationalization. With respect to the GIB and eco-efficient company through the comprehensive method of entrepreneurship, it is observed that just as the company has some advantages in the product, it is also with many threats that are the large companies that are already positioned and that also compete with competitive prices.

Chapter 7 examines whether the countries having high agriculture as well as per capita incomes are responsible for food waste or not using the data on food waste out of export and import channels across the world's selected 20 countries having about 70 per cent contribution to food waste for the period 1971–2019. It finds that food waste is associated with agriculture as well as per capita incomes of the developing and developed countries (DCs), positive for majorly developing countries and negative for majorly DCs. The time series analysis involving food waste, per capita agriculture income and per capita income as the indicators, the study finds that there are long-run relations among them across many countries implying countries having high incomes are responsible for food waste in most cases due to their food preference, although having good infrastructural facilities for preservation of foods.

Chapter 8 inspects the improvement of rice production in Nigeria through the government's policy of the Anchor Borrowers' Programme (ABP) using a cross-sectional survey research design to measure rice production and ABP in Ebonyi State, Nigeria. Applying regression technique, the study reveals the need for the ABP, banks and other lending agencies/programmes to train farmers. It shows that modern farming contributes to rice production in Ebonyi State.

Chapter 9 attempts to explore how a shift toward zero-budget farming from inorganic farming can improve the access of agricultural goods, as predominant component of their export basket, to the markets in DCs, given the rising stringency of sanitary and phytosanitary and environmental standards, as well, coming disproportionately against the exports from developing and less-developed countries (LDCs), including India. With reference to Balassa's the measure of Revealed Comparative Advantage, the chapter points to the dwindling competitive advantage of India's agriculture in global market during the last decade. Given the stringent environment standard facing, disproportionately, the exports from LDCs, zero-budget farming can improve the access of agricultural goods, as predominant component of their export basket, to the markets in DCs.

Chapter 10 mentions that agriculture plays an important role in human life, contributing to the national economy. Though its biggest problems lying in the carbon gas emission during production through fertilizing activities, etc., the study is aimed to determine the most appropriate strategy for carbon emission in this study. As a result of the DEMATEL analysis, it was seen that the most appropriate strategy was effective regulations and auditing.

Chapter 11 points out that sustainable agriculture comprising with environmentally friendly farming methods allows crops to be produced without harming natural systems and preventing adverse impacts on soil, water, biodiversity as well as surrounding or downstream resources and analyses all issues related to mitigating environmental degradation through agricultural sustainability. It helps to take necessary measures for designing and its actual execution. The study constructed a composite index of agricultural sustainability by applying the principal component analysis method where significant differences in agricultural sustainability were found among the states in India. It is also to mention that agricultural sustainability has improved for all states in 2019 compared to 2016.

Chapter 12 aims to examine the interrelationships between crop production, livestock production, transport services and CO_2 emissions from 1998 to 2019. To achieve this goal, the study begins by conducting stationary tests to determine the order of integration for the variables under consideration. Following this, panel unit root tests are employed, and subsequently, panel cointegration tests are conducted to identify any long-term relationships among the selected variables. The findings reveal a significant long-term relationship among crop production, livestock production, transport services and CO_2 emissions.

Chapter 13 investigates agricultural production by sector, convergence and environmental sustainability using West African countries as a case study for the period 2000–2022 using system dynamic panel-data estimation. Based on the findings, it is concluded that there is a divergence effect of agricultural production by sector. The agricultural growth levels of economies are diverging, at a rate of 0.03%, 0.02% and 0.02%, respectively, for the crops subsector, livestock subsector and total agricultural production.

Chapter 14 focuses on the importance of undergraduate and graduate curriculums of future architecture, engineering and construction (AEC) industry professionals to equip them with the skills and knowledge of sustainable agricultural practices and technologies integrated greenhouses, built environment and indoor environment and interior design. For this reason, agricultural policies need to cover food security-related interdisciplinary education and training (e.g. renewable energy-based agriculture integrated built environment) of AEC industry professionals. The study recommends that agricultural policies need to be designed with the contribution of, and considering, AEC industry professionals as they are among the main stakeholders of food security and renewable energy-based agriculture-integrated built environment and how it makes compliance with United Nations Sustainable Development Goals and countries sustainable and resilient development plans can contribute to food security and sustainability.

Chapter 15 examines the technical efficiency of small pond fishery using the stochastic frontier model in the northern region of the state of West Bengal in India taking the data from field surveys in Uttar Dinajpur and Dakshin Dinajpur districts for three months (i.e. from May to June) in the year 2022. The estimated stochastic frontier model revealed that investment in labour, organic fertilizers, fish fingerling and land area could increase the return from fish production. The result shows that technical efficiency ranges from 83 to 100 per cent with a mean efficiency of 94 per cent implying that average fish farmers in the study area are performing below the maximum possible production level by a shortfall of about 6 per cent, though with the issue of sustainability.

The final chapter, that is, Chapter 16, explores Bangladesh's vulnerability to environmental stressors and climate change impacts which significantly affect food availability and agricultural productivity highlighting the pervasive issue of food safety in the country, identifying contaminants such as pesticide residues as a major threat to public health and food security. Then it focuses on the country's huge potential to ensure technology-driven agriculture production that may further strengthen food security and the economy with a special focus on the governance and environmental issues in ensuring sustainable agriculture. It recommends that the Government of Bangladesh needs to adopt a policy that will ensure the national interest and the safety of the citizens inside the country which will help to avoid any devastating food crisis during emergencies and international disputes.

Summary

The studies so far covered up in the book have mainly aimed at examining the economic and environmental implications of agricultural practices in countries and groups. The summary observations are that the countries have experienced rising growth of agricultural sector leading to the improvements in the overall economic activities. There are so many institutional supports from banks, several borrowing programmes, entrepreneurship developments, among others, which have worked as the catalyst for the additional growth progress in the agricultural sector. It has also been observed that the DCs with high per capita income and relatively low dependence on agriculture sector do waste more food due to their preferences over good quality foods compared to the so-called low income countries. Beside the magnificent achievements in growth front, the activities in the sector have led to the environmental degradation although a list of studies have focused on recommending the climate changing issues. Zero budgeting toward inorganic fertilizers usage and full budgeting toward organic farming practices is also recommended. Thus, as the concluding observation, the entire book has revealed the spectacular growth of the agriculture sector during the modern times with some sort of negative externalities in terms of environmental degradation. Both developed and less-developed economies should focus upon sustainable agricultural practices, as has been recommended by the listed research on the topic to have a better future so that inter-generational equities can be maintained.

References

Čolović, V., & Petrović, N. M. (2014). Crop insurance – Risks and models of insurance. *Economics of Agriculture, Institute of Agricultural Economics*, *61*(3), 561–573.

Das, R. C. (2023). Sustainability through total factor productivity growth in agriculture incorporating institutional factors: A post-globalized Indian scenario. *International Journal of Social Ecology and Sustainable Development*, *14*(1), 1–16.

Das, R. C., & Mukherjee, A. (2020). Interplays between methane emission and agricultural output: Time series outcomes for the world's low- to high-income groups. *International Journal of Social Ecology and Sustainable Development*, *11*(4), 56–69.

Das, R. C., Das, A., & Ray, K. (2018). Examining forward and backward linkages between public and private investments: A cross-country analysis. *Review of Market Integration*, *10*(1), 1–31.

Duxbury, J. M. (2001). Long-term yield trends in the rice-wheat cropping system. *Journal of Crop Production*, *3*(2), 27–52.

Enjolras, G., Capitanio, F., & Adinolfi, F. (2012). The demand for crop insurance: Combined approaches for France and Italy. *Agricultural Economics Review*, *13*(1), 5–22.

Gangopadhyay, S. (2004). An alternative to crop insurance. *Economic and Political Weekly*, *39*(44), 4763–4764.

Haddad, B. M. (2005). Ranking the adaptive capacity of nations to climate change when socio-political goals are explicit. *Global Environmental Change*, 15, 165–176.

Hope, C. (2008). Discount rates, equity weights and the social cost of carbon. *Energy Economics*, *30*(3), 1011–1019.

Volenzo, T. E., Odiyo, J. O., & Obiri, J. (2019). Greenhouse gas emissions as sustainability indicators in agricultural sectors' adaptation to climate change: Policy implications. Jàmbá: Journal of Disaster Risk Studies, *11*(1), 576. https://doi.org/10.4102/jamba.v11i1.576

Chapter 2

Change in Productivity of Cereals Cropping and Trade Relationship among BIMSTEC Countries

Subhrabaran Das and Kiran Bhowmik

Department of Economics, Tripura University, India

Abstract

Agriculture is one of the important sectors for the sustainable development of any country. In order to achieve the SDG1 and 2, improvization of productivity and performance in agricultural development is needed. The present study attempts to analyse the change in productivity and cropping pattern among the Bay of Bengal Initiative for Multi-sectoral Technical and Economic Cooperation (BIMSTEC) countries that comprises Bangladesh, Bhutan, India, Myanmar, Nepal, Sri Lanka and Thailand. The study also examines the performance of BIMSTEC countries in exporting the agricultural products. The study is based on the secondary data collected from Food and Agriculture Organization of the United Nations. In accordance with the importance for the development of the country, the study also analyses the export and import of these cereal crops among the BIMSTEC countries. For the empirical investigations the study uses co-efficient of variation, Coppock's instability index and Balassa's index. The study reveals that productivity and cropping pattern of the cereal production has been changing gradually over time in all BIMSTEC countries which show the importance of products substitution. It also reveals that the development cooperation among the BIMSTEC countries through promotion of exports and imports is required to meet their needs.

Keywords: Cereal cropping; change in productivity; Coppock's Instability Index; export and import; BIMSTEC countries

Sustainable Agricultural Practices: Economic and Environmental Implications, 11–26

doi:10.1108/978-1-83608-336-820241002

Introduction

Agriculture is one of the important sectors through which regional cooperation can be done for the sustainable development of the countries. In order to achieve the SDG1 and 2, agricultural development is needed through improvement of productivity and performance. Better productivity and performance in agricultural sector indicate the favourable trade-off relationship with the secondary sector and tertiary sector in any country and as a result the development of that country accelerates. Agricultural development may be deemed as the critical aspect of inclusive growth and proves to be a smooth path for achieving social and economic inclusion (Das, 2021). Bolt (2004) has explained that agricultural growth and rural development have significant potential to achieve inclusive growth among the Asian countries. Growth of agriculture sector is the key aspect for poverty reduction and inclusive growth.

The Bay of Bengal Initiative for Multi-Sectoral Technical and Economic Cooperation (BIMSTEC) is a multilateral regional organization established in 1997 with the aim of accelerating shared growth and cooperation to improve mutual trade, culture and economic development among the adjacent countries in the Bay of Bengal region (viz. Bangladesh, Bhutan, India, Nepal and Sri Lanka from South Asia and two countries – Myanmar and Thailand from South-East Asia). India is the eldest sister among the BIMSTEC nations. BIMSTEC is an opportunity to these countries to build greater regional trust, resilience and sustained efforts for its realization in the field of economic development. Initially, there are six major areas viz. Trade and investment, Transport and communication, Energy, Tourism, Technology and Fishery. Later, other eight priority sectors are added including agricultural and food security.

Agriculture is one of the important priority areas in BIMSTEC countries. There are more than 1.7 billion people which are about 22 per cent of the global population live in BIMSTEC countries. Agriculture is the largest requirement of the common people among these countries. Trading is also an important priority area through which these countries are maintaining their socio-economic development of the region. These countries are maintaining tradable relationship in exporting and importing of agricultural products among each other in order to fulfil SDG1 and SDG 2 and to improve the socio-economic development among them. Trade is an exchange that allows countries to exchange goods and services in view to get different advantages of trade. The trade policy of BIMSTEC countries can become one of the major contributors to the reduction of poverty in the region. Trade gives the opportunities for consumption from greater variety of available goods in these countries. The trade among BIMSTEC countries can help not only to reduce the poverty but also helpful for rapid economic development of these countries. In order to create an enabling environment for rapid economic development through identification and implementation of specific cooperation the BIMSTEC nations are involved among each other in trading, investment and industry, technology, human recourse development, tourism, agriculture, energy and infrastructure and transportation. The global trade especially in agricultural sector can become one of the major contributors to the reduction of poverty and hunger. In agricultural trade, cereals are the main tradable products and play a crucial role in achieving SDG 1 and 2 through reducing poverty and hunger among the BIMSTEC nations. The cereals cropping viz. maize, millet, rice and wheat are vital staple cropping produced in these countries. In India it has observed that the agricultural production in case of food grain was the highest priority but the concept has been changing due to cropping pattern turned against pulses, coarse cereals and sugarcane (Ghosh, 2010). Though the cropping pattern is still dominated by food grain but the non-food grain crops like oilseeds, cotton, sugarcane, etc. have been continuously replacing the food grain crops in India (Ghosh, 2011).

The agricultural export is still an important sector for economic development in the case of developing countries (McPherson, 1966). During 2001–2002, the share of agricultural exports in total exports was 21.47 per cent and it grew at an annual growth rate of nearly 6.53 per cent during 1990s with significant increase in export of rice (Singh et al., 2006). India is expecting export prospects and the agricultural exports have increased. The higher agricultural export of agricultural goods would mean good price of agricultural goods for farmers, positive change regarding good agricultural practices and improving the quality of agricultural goods (Bhatia, 2021).

This chapter examines the change in productivity of cereals cropping and trade relationship among the BIMSTEC countries over the period of 2000–2020. The change in productivity of cereals cropping may vary depending on the environmental conditions, soil, variety and fertilizers. As there is change in productivity of cereals cropping, the countries take care in producing these cereal goods where they have a comparative advantage. The comparative advantage of cereals cropping also provides countries a better marketing power with the help of trade. The study also attempts to examine the trade relationship among BIMSTEC countries in exporting the agricultural products. The study is mainly focused on cereal cropping.

Objectives

This chapter is based on the following objectives:

1. To examine the change in cropping pattern of cereal cropping in India and other BIMSTEC countries.
2. To examine the extent of trade relationship among BIMSTEC countries through the analysis of export and import of the principal crops.

Data Sources and Methodology

The study is based on the secondary data sources for a period of 20 years from 2000–2020 which is collected from the Food and Agriculture Organization (FAO) of the United Nations. The study is concentrated on the principal cereal crops of all BIMSTEC nations. The selected principal cereal crop productions viz. rice, wheat, maize and millet of all BIMSTEC nations have been taken for this study.

In order to examine the change in cropping pattern of cereal cropping, expansionary effect and substitution effect are measured. For a given gross-cropped area the substitution effect is defined as the relative decline in the area of some crops and compensating increase in the area of other substitutable crops. Whereas the expansionary effect of cropped area that is shown with the change in cropping pattern is also discussed more clearly by dint of 'cropped area gross cropped area elasticity (E)' (Venkataraman & Prahladachar, 1980):

$$E = \frac{\%\text{ change in the growth of area of a crop}}{\%\text{ change in gross cropped area}}$$

The substitution effect is shown by using the following formula:

$$\begin{array}{c}\text{Cross elasticity for}\\ \text{substitute croping}\end{array} = \frac{\%\text{ change in the growth of area of a crop}}{\%\text{ change in the area of contracting crop}}$$

The instability and growth in cultivated area and the production of cereal cropping in India and other BIMSTEC countries are done prior to examine the expansionary and substitution effects. In order to construct the instability index of a parameter several methodologies such as moving average, standard deviation, coefficient of variation, etc. can be used (Anjum, 2018). However, in the present study Coppock's instability index (CII) (1962) is used to measure the instability or change in production, yield and its area:

$$\text{CII in per cent} = \text{Antilog}(\sqrt{V(\log)} - 1) \times 100 \quad (1)$$

$$\text{where } V \log = \frac{\sum \left(\log \frac{X_{t+1}}{X_t} - m \right)^2}{n-1}$$

X_t is the area/production/yield in the tth time period, n is the total number of periods, m the arithmetic mean of the difference between the logs of $X_t + 1$ and X_t and $V(\log)$ the logarithmic variance of the series.

The growth rate of cropped area, production and productivity has been estimated by using the following exponential growth model (Veena, 1996):

$$\log Y = \alpha + \beta t + u \quad (2)$$

where Y is a dependent variable viz. area, production, productivity, α is the intercept parameter, β is the growth rate, t is a time variable and u is the stochastic error component, which follows independently identically normal distribution with mean μ and variance σ^2. The above equation is also used to estimate the export and import growth of the cereal production.

In order to examine the performance of the BIMSTEC countries in exporting the agricultural products, Balassa's indices (1965) have been used. The index of revealed comparative advantage (RCA) for country *i* for *k*th agricultural product in the *t*-year can be expressed as

$$\mathrm{RCA}_{ij}^{(t)} = \left[\frac{\frac{x_{ik}^{(t)}}{X_i^{(t)}}}{\frac{x_{wk}^{(t)}}{X_w^{(t)}}} \right]$$

In equation (3), $x_{ik}^{(t)}$ denotes the value of export of *k*th agricultural product by the *i*th country in the *t*-year and $X_j^{(t)}$ represents the value of total export of all agricultural products by the *i*th country in the *t*-year. Similarly, $x_{wk}^{(t)}$ refers to the export of *k*th agricultural product by all countries in the time *t* and $X_w^{(t)}$ indicates the value of export of all agricultural products by all countries in the *t*-year. The range of value of the index of RCA is $0 \le \mathrm{RCA} \le \infty$. A country is said to have comparative advantage in exporting the agricultural crops if RCA is higher than unity. In contrary, a country is said to have comparative disadvantage in exporting the agricultural crops if the value of RCA is lesser than unity. The index of RCA, thus, determines a country's strength of export analysis in agricultural products.

Result and Discussion

Change in Cropping Pattern in India from 2000 to 2020

In order to check the change in cropping pattern of cereal production, area of land used for the particular cereal cropping over time and estimation of cropping elasticity and cropping substitution are important. Table 2.1 shows the expansionary and contraction effect among BIMSTEC.

Table 2.1 reveals the crop area–gross-cropped area elasticity under crops among BIMSTEC nations from 2000 to 2020. The positive elasticity of cropped area shows the expansionary effect while negative elasticity shows the contraction effect in cropped area.

Table 2.1 reveals that the crop area elasticity value is more than one in India, which indicates the crop area expansion for cropping rice (1.576), wheat (12.320) and maize (31.262), whereas negative elasticity is found for the production of millet (−30.174), which indicates that cropped area contraction is found in the case of millet production. In the case of Bangladesh, area elasticity value is positive and more than one for production of maize (19.559) and rice (1.067) and less than zero for the production of millet and wheat. In the case of Bhutan, area contraction is found for all cereal crops. Table 2.1 also reveals that area of millet crop is contracted in all countries except Nepal and Thailand; area of wheat production is contracted in Bangladesh and Myanmar, rice is contracted only in Nepal and maize is contracted only in Thailand. In Thailand, millet is not cultivated and in Sri Lanka no wheat production is found.

Table 2.1. Expansionary, Contraction and Substitution Effect in Cropped Area among BIMSTEC from 2000–2001 to 2020–2021.

Expansionary and Contraction Effect

Country/crops	**Maize**	**Millet**	**Rice**	**Wheat**
Bangladesh	19.559 ($E > 1$)	−24.182 ($E < 0$)	1.067 ($E > 1$)	−29.751 ($E < 0$)
Bhutan	−11.3 ($E < 0$)	−31.987 ($E < 0$)	−6.466 ($E < 0$)	−24.491 ($E < 0$)
India	31.262 ($E > 1$)	−30.174 ($E < 0$)	1.576 ($E > 1$)	12.320 ($E > 1$)
Myanmar	3.561 ($E > 1$)	−3.807 ($E < 0$)	0.315 ($0 < E < 1$)	−2.088 ($E < 0$)
Nepal	5.486 ($E > 1$)	0.862 ($0 < E < 1$)	−1.192 ($E < 0$)	3.063 ($E > 1$)
Sri Lanka	3.858 ($E > 1$)	−37.276 ($E < 0$)	1.338 ($E > 1$)	–
Thailand	−0.7725 ($E < 0$)	–	0.355 ($0 < E < 1$)	0.444 ($0 < E < 1$)

Substitution Effect (Crop Area Substitution)

Country/Crops	**Contraction**	**Maize**	**Millet**	**Rice**	**Wheat**
India	Millet	−1.036	–	−0.052	−0.408
Sri Lanka		−0.01	–	−0.003	–
Bangladesh	Millet	−0.808	–	−0.044	–
	Wheat	−0.657	–	−0.035	–
Myanmar	Millet	−0.935	–	−0.082	–
	Wheat	−1.705	–	−0.15	–
Nepal	Rice	−4.602	−0.723	–	−2.569
Thailand	Maize	–	–	−0.46	

Source: Calculated from data https://desagri.gov.in.
Note: Parentheses indicate the nature of elasticity showing the expansionary and contraction effect.

The substitution effect shows the change in substitute cropping due to changes in cropping. The substitution effect is the cross-crop area elasticity that is informed by the change in the area of one crop with respect to the change in area of other cropping. Table 2.2 shows the substitution effect of the crop whose area is contracted over time among the BIMSTEC countries. It is also observed that cross-crop area elasticity is high between maize and millet in India (−10.036), Sri Lanka (−0.01) and Myanmar (−0.935) compared to other crops in these countries. It is also found that maize is the better substitute crop than other crops as the absolute value of cross elasticity of maize and millet is higher than other crops in these three countries. Maize is also a better substitute of wheat in Bangladesh (−0.657) and Myanmar (−1.705), and maize and wheat are also a better substitute than rice in Nepal. In Thailand, wheat is the better substitute of maize as it is contracted in Thailand.

Instability of Area, Production and Yield among BIMSTEC Countries

The instability that measures through log variance method and the index with higher numerical value represent greater instability and the lowest numerical value getting by dint of CII calculation.

Table 2.2 shows the analysis of instability in area, production and yield for different crops among BIMSTEC countries. Table 2.2 also reveals that the maximum instability in the area for the crop maize (0.11) in the country Bangladesh among BIMSTEC countries, which is followed by Sri Lanka (0.03), Bhutan (0.02), Myanmar (0.01) and Thailand (0.01). The index value of area instability is minimum and negligible for maize cropping in India and Nepal (0.0007). The area instability is highest (0.54) for millet cropping in Sri Lanka among BIMSTEC nations, which is followed by Bangladesh (0.11), Bhutan (0.05), Myanmar (0.02) and India (0.01). It is observed that the area instability is lowest for the crop millet in Nepal (0.0012). The area instability for paddy is observed very less in Bangladesh, India, Myanmar, Nepal and Thailand. In Sri Lanka, the area instability for paddy cropping is highest (0.02). The area instability of rice is observed as 0.01 in Bhutan.

The area instability of wheat is observed as highest in India and Thailand (0.56), which is followed by Bhutan (0.08), Myanmar and Bangladesh (0.02). The instability of area for wheat is reported lowest in Nepal (0.002). The instability in the production of maize is highest in Myanmar followed by Sri Lanka, Bangladesh, India, Nepal, Thailand and Bhutan; whereas in case of millet production the instability is highest in Bangladesh followed by Myanmar, Bhutan, Sri Lanka, Nepal and India. The instability in rice production is highest in Myanmar followed by Bangladesh, Sri Lanka, India, Thailand, Nepal and India. For production of wheat, the instability is highest in Thailand followed by Bangladesh, Nepal, Bhutan, Myanmar and India.

Among these four principal cereal crops, the instability in rice production is lesser than other crops in these BIMSTEC countries. Thailand reports highest instability in the production in the production of wheat and lowest instability in the production of maize. In Bangladesh, the highest instability is shown in the production of maize and highest and lowest in rice production. In India,

Table 2.2. Instability of Area, Production and Yield of the Principal Crops among BIMSTEC Countries During 2000–2001 to 2020–2021.

Country	Maize			Millet			Rice			Wheat		
	Area	Production	Yield	Area	Production	Yield	Area	Production	Yield	Area	Production	Yield
Bangladesh	0.11	83.04	96.52	0.11	80.04	47.9	0	55.06	52.4	0.02	82.02	51.66
Bhutan	0.02	48.7	56.89	0.05	61.41	50.97	0.01	46.21	51.7	0.08	67.07	50.89
India	0.0007	62.82	52.61	0.01	43.93	51.98	0	50.94	48.53	0.56	57.67	50.01
Myanmar	0.01	115.17	62.01	0.02	64.4	56.99	0	56.12	47.45	0.02	62.23	54.37
Nepal	0.0007	56.48	45.14	0.0012	53.61	39.73	0	49.59	46.28	0.002	69.72	50.83
Sir Lanka	0.03	91.86	64.98	0.54	56.82	47.22	0.02	52.97	46.25	–	–	–
Thailand	0.01	49.13	49.58	–	–	–	0	50.34	45.18	0.56	83.87	89.76

Source: Author's calculation is based on data from Food and Agricultural Organization (www.fao.org).

the instability is highest in maize production and lowest in millet production. The instability in yield for maize production is higher than other cereals in all BIMSTEC countries except Thailand. In Thailand, instability in yield is highest for the production of wheat. In Bangladesh, the least value of instability in yield is found for the millet production and the highest is found in maize production. In Nepal, the lowest instability in yield is found in millet production.

There is a close association between area and production of cereals among different BIMSTEC countries because the higher instability of area indicates higher instability of production for cropping maize, millet, rice and wheat. In the case of maize production, the degree of instability in area and production is highest in Bangladesh and lowest in Nepal; however, yield instability is highest in Bhutan and lowest in Nepal. The highest degree of instability in production of maize is due to non-availability of quality seeds (Manjruzzaman et al., 2009). In the case of millet production, the degree of instability in area and production is highest in Sri Lanka; however, it is lowest in Nepal. The yield instability of millet production is highest in Bangladesh and lowest in Nepal. In the case of rice cropping, the instability in area and production is highest in Sri Lanka; however, it is lowest in Bangladesh and Myanmar. The area instability is also lowest in Bangladesh, India, Thailand and Nepal. The yield instability for rice production is highest in Bhutan and lowest in Bangladesh, Myanmar and Thailand. In the case of wheat production, the degree of instability in area, production and yield is highest in India and Thailand. The degree of instability in area and production for wheat is lowest in Nepal.

Table 2.3 reveals the growth of area and production of all BIMSTEC countries under different cereal production over the last 50 years. From the table, it is observed that the growth in area of land for all cereals production is positive and significant over the last 50 years. In Bangladesh, the growth rate of cultivated land area (0.13) is more than that of other cereals production that indicates that Bangladesh concentrates more on maize production. In Bhutan, the growth rate in area of land is negative for all cereals production in the country over the last 50 years, which implies that the people belong to Bhutan do not concentrate on cereal production. In India, the same and positive growth (0.01) is observed for both cereals, maize and wheat, but the negative growth is found for millet that implies that the country has given less importance to millet production. The growth of land for maize production (0.04) is highest and positively significant in Myanmar, which is followed by paddy and millet (0.01). An equal area of growth (0.02) over time is found for the cereals – maize, millet and wheat – in Nepal. Three major cereal crops such as maize, rice and millet are produced in Sri Lanka. It is observed that the area growth of millet is negative, which indicates that millet is produced less compared to the other two cereals. Thailand is producing three major cereals, maize, paddy and wheat, and it is found that the area of growth for wheat (0.18) is highest compared to other crops.

Table 2.3 shows the growth of production for the major crops among the BIMSTEC countries during the last 50 years. From Table 2.3, it is observed that there is a positive growth in crops such as maize, paddy and wheat in Bangladesh, whereas negative growth (−0.04) is found for millet production. In Bhutan, negative growth is found for crops millet and wheat and positive growth for paddy (0.01). In India, positive and significant growth is found for all crops and the

Table 2.3. Growth Rates of Area and Production for Major Crops among BIMSTEC Countries During 1971–2020.

Countries	Maize		Millet		Rice		Wheat	
	Area	**Production**	**Area**	**Production**	**Area**	**Production**	**Area**	**Production**
Bangladesh	0.13* (14.142)	0.19* (14.9)	−0.04* (−8.44)	−0.04* (−8.76)	0.00* (11.28)	0.03* (45.55)	0.02* (3.17)	0.03* (5.65)
Bhutan	−0.02* (−11.35)	0 (−0.10)	−0.03* (−7.24)	−0.02* (−4.36)	−0.01* (−5.59)	0.01* (3.58)	−0.04* (−7.83)	−0.02* (−4.01)
India	0.01* (15.0)	0.04* (28.4)	−0.02* (−27.81)	0.01* (3.52)	0.00* (10.14)	0.02* (25.02)	0.01* (23.48)	0.03* (28.22)
Myanmar	0.04* (24.77)	0.08* (25.72)	0.01* (4.68)	0.03* (8.77)	0.01* (12.01)	0.03* (19.7)	0 (0.25)	0.02* (4.85)
Nepal	0.02* (17.14)	0.03* (28.84)	0.02* (14.74)	0.02* (14.3)	0.00* (8.71)	0.02* (19.37)	0.02* (14.39)	0.04* (31.56)
Sir Lanka	0.02* (7.43)	0.04* (11.67)	−0.04* (−14.91)	−0.03* (−11.77)	0.01* (8.9)	0.03* (17.81)	–	–
Thailand	0 (2.38)	0.02* (8.46)	–	–	0.01* (10.33)	0.02* (19.78)	0.18* (10.38)	0 18* (11.42)

Source: The author's calculation is based on data from Food and Agricultural Organization (www.fao.org).
Notes: Digits in parentheses are the corresponding *t*-values and *indicates 1 per cent level of significance, **indicates 5 per cent level of significance and ***indicates 10 per cent level of significance.

growth rate is highest for maize (0.04) compared to other crops such as millet (0.01), paddy (0.02) and wheat (0.03). In India, the demand for wheat is very high and it is considered as the second most important food crop after rice consummation. The consumption of wheat has increased to 104.08 million metric ton during 2022–2023 from 95.4 million metric ton during 2019–2020,[1] and India is considered as the second largest wheat-producing country in the world. The wheat production has increased to around 78.4 million tons in 2007–2008 from a mere 12.3 million tons in 1964–1965 and the country has been sustaining the wheat productivity as essential to provide food security to the population of India (Yadav et al., 2008). A positive and significant growth is also found for all crops in Myanmar, where the growth is highest for the production of maize (0.08) and lowest for wheat (0.02). In Nepal, the highest growth is found for wheat production (0.04) followed by maize (0.03) where millet and paddy have the growth production of 0.02. In Sri Lanka, negative growth in production is observed for millet (−0.03) but positive growth for maize (0.04) followed by paddy (0.03). In Thailand, highest positive and significant growth of production is found for wheat (0.18) and the production growth is same (0.02) for maize and paddy. Crop-wise highest significant and positive growth rate for maize (0.19) is found over time in Bangladesh among BIMSTEC countries, whereas the growth of wheat production is highest in Thailand (0.18). This indicates that Bangladesh is focusing more on producing maize and Thailand is on wheat; it is also found from the highest growth rate of area for cropping both maize (in Bangladesh) and wheat (in Thailand).

Table 2.3 also shows the highest growth of maize in area, yield and production under the crop maize in Bangladesh, whereas the crop wheat is reported to be the highest growth in area, yield and production in Thailand. Nepal reports millet as the highest in the area, whereas India and Myanmar reveal millet as highest in the yield and the Myanmar highest in millet production. The highest area growth for crop rice is in Myanmar, Sri Lanka and Thailand, and the yield growth reports rice as the highest growth under crop rice in Bangladesh, Bhutan, India, Myanmar and Sri Lanka, whereas the production of crop rice reveals the highest growth in Bangladesh, Myanmar and Sri Lanka.

Bhutan has reported the lowest growth in area and production under the crop maize where the Nepal's growth in yield under the crop maize indicates the lowest. Though Bangladesh reports as lowest growth in area, yield and production under the crop millet, Sri Lanka and Nepal also show lowest growth in area and yield under the crop millet. Bhutan reports lowest growth in area and production under the crop rice but Nepal and Thailand reveal lowest growth in yield for the same crop rice. Bangladesh also shows lowest growth in area, yield and production but Bhutan, India and Myanmar show lowest growth under the crop wheat during present discussion on the basis of data collected from FAO from 1971 to 2020.

Table 2.4 shows the export and import growth of different crops among the BIMSTEC countries over the last 50 years. The export growth of maize is

[1]*Source:* https://www.statista.com/statistics/1310589/india-wheat-consumption-volume/ accessed on 10 February 2024.

Table 2.4. Export and Import Growth Rates of Some Major Crops Among BIMSTEC Countries During 1971–2020.

Country	Export				Import			
	Maize	Millet	Rice	Wheat	Maize	Millet	Rice	Wheat
Bangladesh	0.04** (−2.26)	0.01* (−3.38)	0.10* (−4.9)	0.01 (−1.49)	0.36* (−17.63)	0.06* (−5.01)	0.22* (−6.42)	0.03* (−6.92)
Bhutan	0.01 (−0.57)	0 (−0.24)	0 (−0.4)	−0.28 (−0.17)	0.25* (−12.91)	0.01 (−0.52)	0.11* (−5.96)	0.13* (−4.82)
India	0.30* (−10.33)	0.26* (−12.29)	0.27* (−9.29)	0.06*** (−1.71)	0.06 (−1.57)	0.05* (−2.91)	−0.23* (−7.13)	−0.09* (−2.93)
Myanmar	0.08* (−9.46)	0.14* (−5.47)	0.13* (−4.38)	0.07* (−4)	0.23* (−12.52)	0.04* (−2.66)	0.06* (−3.29)	0.31* (−8.6)
Nepal	−0.09* (−2.79)	0.06* −2.87	−0.24* (−7.33)	0.04 (−1.46)	0.31* (−12.09)	0.24* (−10.25)	0.24* (−7.16)	0.11* (−3.69)
Sri Lanka	0.02 (−0.89)	−0.06* (−2.84)	0.11* (−13.56)	0.21* (−8.98)	0.16* (−5.93)	0.23* (−13.78)	0.04 (−1.51)	0.05* (−10.38)
Thailand	−0.06* (−5.61)	0.14* (−5.79)	−0.05*** (−1.67)	0.09* (−5.12)	0.17* (−10.41)	0.15* (−8.75)	0.07* (−2.84)	0.08* (−28.58)

Source: Author's calculation is based on data from Food and Agricultural Organization (www.fao.org).

Notes: The value in the parentheses indicates the corresponding *t*-values and * indicates significant at 1 per cent level of significance, ** indicates 5 per cent level of significance and *** indicates 10 per cent level of significance.

negatively significant in Nepal (−0.09) and Thailand (−0.06). The export growth rate of millet is positive and highest in India (0.26) which is followed by Myanmar and Thailand (0.14), Nepal (0.06) and Bangladesh (0.01).

The export growth of millet is negative and significant in Bhutan (−0.00) and Sri Lanka (−0.06). The export growth of rice is highest and significant in India (0.27), which is followed by Myanmar (0.13), Sri Lanka (0.11) and Bangladesh (0.10). The export growth rate of rice is positive but minimal in Bhutan (0.00). The rice export is negatively significant in Nepal (−0.24) and Thailand (−0.05). The export growth of wheat is positively significant and highest in Sri Lanka (0.21) which is followed by Thailand (0.09), Myanmar (0.07), India (0.06) and Nepal (0.01). The export growth rate is negative and significant in the case of wheat in Bhutan (−0.28).

It is also observed that import growth of rice is highest and significant in Bangladesh (0.36) which is followed by Nepal (0.31), Bhutan (0.25), Myanmar (0.23), Thailand (0.17), Sri Lanka (0.16) and India (0.06). The millet import growth is highest and significant in Nepal (0.24) which is followed by Sri Lanka (0.23), Thailand (0.15), Bangladesh (0.06), India (0.05), Myanmar (0.04) and Bhutan (0.01). The import growth of rice is highest and significant in Nepal (0.24), Bangladesh (0.22), Bhutan (0.11), Thailand (0.07) and Myanmar (0.06) and the import growth is negatively significant in India (−0.23). The table shows that the import growth rate of wheat is positive for all the BIMSTEC countries besides India. The import growth of wheat is highest and significant in Myanmar (0.31), which is followed by Bhutan (0.13), Nepal (0.11), Thailand (0.08), Sri Lanka (0.05) and Bangladesh (0.03). The negative import growth of wheat is observed in India (−0.09).

The export and import are in compliance with highest and lowest growth under cereal crops maize, millet, rice and wheat. India reports highest export of the crops, maize, millet and rice, whereas Sri Lanka exports wheat as highest among the BIMSTEC countries. Bangladesh imports more maize and rice, Nepal imports more millet crops and Myanmar imports wheat crops mostly among the BIMSTEC countries. Bhutan reports lowest export in the crops such as maize, millet, rice and wheat, but India is the lowest importer of the crops such as maize, rice and wheat. Bhutan also imports less millet among BIMSTEC countries.

Progressive growth of this index over time shows improvement in the competitive position of a good in the world or in a specific region. Table 2.5 shows the mean of RCA of agricultural crops mainly rice, maize, millet and wheat among BIMSTEC countries during the course of years 2000–2021. India has been observing the existence of RCA in case of rice (15.439) and millet (7.313). As RCA value is more than zero but less than one in case of wheat (0.95) in India, it means that India has lack of comparative advantage in exporting the crop wheat. The country has no superiority or inferiority over other countries in exporting good maize (1.03) because the value of RCA is one in case of exporting maize. Bangladesh has not yet revealed the comparative advantage in exporting the agricultural crops millet, wheat (0.002), rice (0.276) and maize (0.017).

Sri Lanka doesn't have the comparative advantage in exporting the agricultural crops maize (0.002), rice (0.403), millet (0.07) and wheat (0.07). Thailand

Table 2.5. Mean of RCA of Agricultural Crops among BIMSTEC During the Period 2000–2021.

Country	Rice	Maize	Millet	Wheat
India	15.439	1.032	7.313	0.950
Bangladesh	0.276	0.017	0.000	0.002
Sri Lanka	0.403	0.002	0.073	0.070
Thailand	19.064	0.421	0.487	0.000
Nepal	0.180	0.000	1.989	0.000
Bhutan	0.234	0.011	0.000	0.000
Myanmar	33.005	10.737	20.957	0.062

Source: Author's calculation is based on data from Food and Agricultural Organization (www.fao.org).

does indicate RCA in exporting rice (19.06) but this country shows lack of RCA in exporting agricultural crops maize (0.42) and millet (0.49). Thailand reports the non-existence of RCA in exporting the agricultural crop wheat (0.00). Nepal doesn't have the RCA in exporting the agricultural crops maize and wheat (0.00) though it has revealed a comparative advantage in exporting millet (1.99). Bhutan has no comparative advantage in exporting the agricultural crops millet and wheat (0.00), rice (0.234) and maize (0.011). Myanmar has revealed the comparative advantage in exporting rice (33.005), maize (10.737) and millet (20.957) though it has a comparative disadvantage in exporting the wheat (0.062).

Conclusions and Policy Implications

From the overall analysis of the cropping pattern changes in terms of 'instability of area, yield and production among BIMSTEC countries' it shows that Bangladesh and Thailand have the highest instability in area, yield and production under the crops maize and wheat, whereas India reports to have lowest instability in the area, yield and production under the crop wheat. India also needs to focus on that with the implementation of different schemes for the benefit of farmers. The analysis also shows the highest growth in area, yield and production under the crop maize in Bangladesh, whereas the crop wheat is reported to be highest growth in area, yield and production in Thailand. India reports to be highest exporting country under the crops maize, millet and rice but Sri Lanka exports wheat as highest. The positive results under exporting of agricultural cropping might be the outcome of government-sponsored schemes launched by central government. Agriculture sector plays a very important role in both economic development and nation building. India is globally in the forefront of development of agriculture. The goal is to achieve doubling of export by 2022.

However, more than 86 per cent of farmers in the country are small and marginal. There is a need to facilitate our farmers with access to improved technology, credit, better input and more markets to incentivize them to produce better quality commodity. For this, aggregation of small, marginal and landless farmers into FPOs will help enhance economic strength and market linkages of farmers for enhancing their income. Keeping this in mind, Government of India has launched a new Central Sector Scheme titled 'Formation and Promotion of 10,000 Farmer Producer Organizations (FPOs)' with a clear strategy and committed resources to form and promote 10,000 new FPOs in the country with budgetary provision of Rs. 6,865 crores.[2]

References

Anjum, S. (2018). Growth and instability analysis in Indian agriculture. *International Journal of Multidisciplinary Research and Development*, *5*(11), 119–125.

Annual Report (2020–2021). Department of Agriculture, Cooperation & Farmers' Welfare, Ministry of Agriculture & Farmers' Welfare, Government of India, www.agricoop.nic.in

Bhatia, J. K. (2021). Export-import performance of micro agricultural commodities in India. *Economic Affairs*, *66*, 117–126.

Bolt, R. (2004). *Accelerating agriculture and rural development for inclusive growth: Policy implications for developing Asia* (pp. 1–9). ERD Policy Brief No. 29, Asian Development Bank.

Coppock, J. D. (1962). *International economic instability: The experience after World War II*. McGraw Hill Book Company.

Das, S. (2022). Impact of globalization on Sustainable Development Goals (SDGs): A study of the State Tripura. In R. C. Das (Ed.), *Globalization, income distribution and sustainable development: A theoretical and empirical investigation* (pp. 311–324). Emerald Publishing Limited.

Das, S. (2021). Determining optimum size of government intervention for sustainable development of agricultural farming in India. In R.C. Das (Ed.), *Optimum size of government intervention emerging economies and their challenges* (pp. 189–203). Routledge.

FAO (2021). Agricultural trade & policy responses during the first wave of the COVID-19 pandemic in 2020. Rome.

Ghosh, B. K. (2010). Essence of crop diversification: A study of West Bengal agriculture during 1970-1971 to 2004-2005. *Journal of Development and Agricultural Economics*, *2*(11), 368–381.

Ghosh, B. K. (2011). Determinants of the change in cropping pattern in India: 1970-71 to 2006-07, *The Bangladesh Development Studies*, *34*, 109–120.

McPherson, W. W. (1966). Role of agricultural trade in economic development. *Journal of Farm Economics*, *48*(2), 354–336.

Ministry of Agriculture & Farmers Welfare (2021). In Proceedings of the *8th meeting of agricultural experts of BIMSTEC countries held*. Media release. Accessed 31 August 2021. https://pib.gov.in/PressReleasePage.aspx?PRID=1750834

Moniruzzaman, M. S. R. S., Rahman, M. S., Karim, M. K., & Alam, Q. M. (2009). Agro-economic analysis of maize production in Bangladesh: A farm level study. *Bangladesh Journal of Agricultural Research*, *34*(1), 15–24.

[2]*Source*: Central Sector Scheme 'Formation and Promotion of 10,000 new Farmer Producer Organizations (FPOs)' of Rs. 6,865 crores at https://pib.gov.in/Pressreleaseshare.aspx?PRID=1696547.

RIS (2011). Ministry of Development of North East Region. Expansion of North East India's trade and investment with Bangladesh and Myanmar. https://mdoner.gov.in/contentimages/files/RIS_Border_Trade_Report_final.pdf

Singh, N. P., Kumar, R., & Singh, R. P. (2006). Diversification of Indian agriculture: Composition, determinants and trade implications. *Agricultural Economics Research Review*, *19*, 23–36.

Veena, V. M. (1996). Growth dimensions of horticulture in Karnataka – An econometric analysis, Ph.D. Thesis, University of Agricultural Sciences, Dharwad, India.

Venkataraman, L. S., & Prahladachar, M. (1980). Growth rates and cropping pattern changes in agriculture in six states: 1950 to 1975. *Indian Journal of Agricultural Economics*, *35*(2), 71–84.

Venkatesh, P., Sangeetha, V., & Singh, P. (2016). Relationship between food production and consumption diversity in India – Empirical evidences from cross section analysis. *Agricultural Economics Research Review*, *29*, 139–148.

Verma, M. R., Datta, K. K., Mandal, S., & Tripathi, A. K. (2007). Diversification of food production and consumption patterns in India. *Journal of Agricultural & Food Information*, *8*(3), 87–100.

Yadav, V. K., De, D., Jirli, B., Kumar, S., & Prakash, S. (2008). Sustainability of scientific wheat cultivation practices in Bihar and Haryana. *Global Communication*, *1*(2), 148–154.

Chapter 3

Natural Resource Evaluation and Climate Change: Analysis of Indian Sundarbans

Nilendu Chatterjee[a] *and Tonmoy Chatterjee*[b]

[a]*Department of Economics, Bankim Sardar College, West Bengal, India*
[b]*Department of Economics, Bhairab Ganguly College, Kolkata, India*

Abstract

UNESCO heritage Sundarbans is perhaps the largest island where people are solely dependent upon natural resources. Indian part of Sundarban covers around 4,263 sq. km. area and mostly situated in the district of South 24 Parganas of West Bengal. Here, people are mainly dependent on agriculture and fishing activities. Since forestry is reserved and people are allowed to collect the non-timber forest products only, the burden of livelihood earning comes on both fisheries and agricultural activities. The mechanization of agriculture and fishery, both of which are costly, have kept on swinging accordingly. People's perception for the sustainable use cum development of these two resources, thus, becomes immensely crucial. Here, we have applied contingent valuation method to know about the people's perception or willingness for saving these resources. The study is based on primary field survey conducted in late 2023 in two different parts of Sundarbans – both of which are dependent on both the resources. We find that people are more willing towards protection of fishery than the agricultural, probably because of its high market value and business opportunities than agricultural products. Again, climate change and frequent storms are more devastating on agricultural products

Sustainable Agricultural Practices: Economic and Environmental Implications, 27–41

doi:10.1108/978-1-83608-336-820241003

than fishery. Such an analysis, in the era of climate change, is not only significant but also unique.

Keywords: Agriculture-Fishery valuation; climate change; contingent valuation method; people's perception; Indian Sundarbans

Introduction

Natural resources play a dual role on environmental protection as well as livelihood generation. Socio-economic-political aspects of lives of resource-dependent communities are directly influenced by natural resources. Hence, conservation, development-cum-efficient use of natural resources are a necessity for sustainable development.

The state of West Bengal is unique in the sense that it is blessed with all forms of climatic zones – beginning from the mangrove-blessed Sundarbans to the dry-deciduous drylands in the western part of the state and extending up to the Himalayas of north Bengal. The state has all forms of agricultural regions and unique climatic conditions. Owing to the nature of the climate, every region has its own natural resource base with dependent communities relying on those resources for their livelihood earnings. Sundarbans is not only unique for its mangrove forestry but also for its huge fishery resources. Even if the proximity to the Bay of Bengal along with the existence of mangroves make the soil saline in Sundarbans, people of Sundarbans do rely on agriculture for sustenance.

Although proportionately larger section of Sundarbans belongs to Bangladesh, the part that belongs to India is also quite large and natural resources bear utmost impact on the lives of the people along with influencing their livelihood earnings. Huge natural resource base of forest products and agricultural production and fisheries along with tourism form the base of livelihood earnings for the people of these areas. Even if this place has a huge forest base and people are forest-dependent, they do not have access to all forms of forest products since the forest products are very much controlled by the forest department and joint forest management committees. People have access to the non-timber forest products (NTFPs) but that is also not easy to get because of frequent attacks from the Royal Bengal Tiger as well as severe effects of climate change – both of which have been very severe in these regions over the last decade.

Climate change, which is a global phenomenon, has had its devastating effect on Indian Sundarbans by means of several storm surges in the last couple of decades and also by increasing the average temperature of the ocean. Sundarbans is the homeland to over 4.5 million people and it is situated near the Bay of Bengal. Here people are heavily dependent on every form of natural resources, but agriculture and fishery hold the prime two positions in this regard.[1] It is

[1]Even though there is considerable dependency on forestry that dependency is seasonal and not widely spread because of non-availability of NTFPs in every parts of

important to understand how climate change has been casting its devastating spells in Sundarbans over the last decade. It is the cyclones that have been more severe and frequent in these parts, of late, and therefore causing rises in the sea level in the Bay of Bengal along with rise in temperature and thus causing climate change.

In the Indian sub-continent both the east and west coast lines are considered as the zones of cyclogenesis, which implies that both the Bay of Bengal as well as the Arabian Sea lines can be considered as cyclone-prone areas (NCRMP 2011). Existing findings suggest that the region of Bay of Bengal is far more cyclone-prone than the Arabian region and the result is the suffering for the states of West Bengal, Orissa, Andhra Pradesh and Tamil Nadu (Singh & Roxy, 2022).[2] This region is especially vulnerable to tropical cyclones because of its location, which lies along a warm ocean current and close proximity to several low-pressure systems. Because of its position in the northern part of this coast line, Sundarbans have been suffering, quite a lot, by storms. In the last three decades or so, there have been an increase in the severity as well as frequency in the storms in these parts. There are more than 42 severe cyclones in the Indian Sundarbans in the last 100 years that have increased the climate-related vulnerability of the 4.5 million people living here. In the last 15 years there have been around eight to nine severe storms that have caused immense suffering for the people of this region.[3] These storms varied from 85 to 270 km/h and caused estimated destruction of above 2,500–2,700 million dollars. It is a fact that flood is a common phenomenon in the villages of Sundarbans during monsoon but mangroves have given protection to millions of people living here; yet such devastating natural disasters have been increasing because of climate change.

The above argument makes it vivid in the presence of severe climate change. Because of risky, insufficient as well as non-sustainable income from forest resources, agriculture and fishery have become two prime sources for the survival of people of Indian Sundarbans. In this study, we have focused on the extraction, upgradation, development and evaluation of these two resources in our study area, that is, selected areas of Indian Sundarbans, by application of contingent valuation method (CVM) which is not only very popular among the environmental economists but also a very widely acceptable method for evaluation of environmental resources. Results of our study bear meaningful implications and have strong impact on the sustainable living of the people of these areas because a high eagerness or desire to contribute for preservation

---

the Sundarbans. What is interesting is that this forest dependency is not good enough for survival since the products available from forestry do not help in surviving the whole year. Hence, dependency on agriculture becomes a natural choice. There are families that have dependency on both forestry and agriculture but among the two, agriculture dominates.

[2]Especially the coastal belts of these regions suffer mainly.

[3]Amphaan, Fani, Yaas and Aila are few of these devastating ones, with Amphan the severe most followed by Fani, when considered speed as well as valuation of destruction caused.

cum sustainable usage of these resources in Indian Sundarbans means that the stakeholders are sustainably aware about the necessity of conserving these resources along with development in their production process, as their livelihood is dependent upon these resources.

Sundarbans is one such place that heavily depend on these two resources but whether the people are really aware or not about the uses of these resources needs to be queried. Our sample size for both agriculture and fishery is 500 covering two important blocks of Sundarbans Hence, our study covers two blocks of Sundarbans, namely Basanti and Namkhana – both of which are dependent on fishery as well as agriculture for survival. So, such a large sample base is expected to reflect the actual scenario of these two resources in resource-dependent Sundarbans. Few works have emphasized the valuation of forestry and fishery Sundarbans (Chatterjee, 2019) but none of them have covered such huge areas and such large number of households. Valuation of these resources is important because severe climate change has forced the people of these regions to go outside the district or even state in search of livelihood (Chatterjee, 2023). People have been changing the use of their lands depending on the effects of climate change – they use land for both agricultural purpose as well as for fishing – depending on profitability as well as weather conditions (Chatterjee, 2023). The salinity of soil prevents multiple cropping in these regions and in the slack season or non-farming season people use their land for the purpose of fish cultivation.[4] All such things demand greater significance of our work. What effects of climate change, global warming and such storms have on the lives of common people and their perception about conservation and sustainable improvement of agriculture demands more priority and significance in research.

During our pilot survey and field survey, we have seen that people are very much aware about development of agriculture as they understand that it generates livelihood. Equal awareness was seen in case of inland fishery as well as open access fishing. People who are cultivating fish, having their own fishery, are aware about the sustainable uses of fishery and the negative impact climate change has on fish cultivation and its growth. But not all fishermen of open access fishing in the Bay of Bengal are aware about the sustainable fishing.

What we felt during our survey is that earning from fishery is much higher than that from agriculture. Hence, through pilot survey and having discussion with the panchayats, farmers and Fishermen Association, we designed our bids accordingly. The bids for fishery are kept higher deliberately than that of agriculture. But whether people possess any willingness in themselves to donate is the prime aspect of this study. Even if there are few studies in the field of valuation of resources of Indian Sundarbans (Chatterjee, 2019; Gupta, 2005), there is no work involving the valuation of agriculture and fishery in a single study; perhaps that has worked as the prime motivation behind this study.

---

[4]Even if it is a fact that open access fishery is the prime form of fishing in these areas, in the last decade or so, inland fishing has gone up significantly and become a source of livelihood.

Valuation of Agriculture: Methodology and Results

A mix of different sampling procedures has been followed here – firstly, for the selection of villages we have followed stratified sampling and then again for the selection of households we have used a mix of stratified and random sampling. Hence, both forms of sampling have been followed but not in full but in parts. Our sampling size for the evaluation of agriculture is 500. This survey on the households has been carried out in Basanti and Nankhana blocks, covering 250 households from each of the two blocks. Here, we would also like to mention the fact that during our survey, we have found 100% responses from the respondents. Hence, our response rate is significantly high though it is very much unconventional but can be considered good for the developing regions. We consider the high rate of response as a benefit of applying 'face to face' interview process following the recommendations by the NOAA panel in 1993.

Our application of CVM is dependent upon close-ended bidding procedure of dichotomous choice. For following such procedures, it becomes immensely important to decide the amount of bid first and then the procedure to offer it. The bids we have considered for valuation of agriculture of Sundarbans have been decided through pilot surveys and after discussing with the members of the Panchayat.[5] Hence, through pilot survey, we decided our bids which are Rs. 15, Rs. 25, Rs. 40, Rs. 50, Rs. 65 and Rs. 75 per month for the sake of sustainable use cum upgradation of agriculture. Upgradation involves all forms of mechanization along with marketing. Next, the important thing was to find out the valid respondents and a way to offer the bids to the respondents by maintaining randomness. For this purpose, we have categorized the six bids into three groups and accordingly respondents were also categorized into those three groups. These groups, according to the offerings of bids are low, medium and high. Bids of Rs. 15 and Rs. 25 are in the *lower* bidding group, Rs. 40 and Rs. 50 are in the *middle* bidding group and Rs. 65 and Rs. 75 are in the *higher* bidding group. Since we have used close-ended single-bounded dichotomous choice model in our study, offerings of bids to the respondents were done by *dual dice-throwing process*. This process not only maintained the randomness of offering a certain bid to the respondent but also the randomness in categorizing of a specific respondent to a specific bidding group. Hence, we maintained randomness in all the possible stages of the offerings of bids.

A dice was first thrown to each and every respondent or household.[6] When the corresponding result from dice-throwing was either 1 or 6, the corresponding respondent was categorized in the 'low bidding group'. For this specific respondent, the offered bids were either Rs. 15 or Rs. 25, depending on the second dice throw. For deciding which of the two bids were to be offered, again a dice was thrown which can be considered as dice throw of round two. For the second

[5]We wanted to find a better idea about the socio-economic well-being of the people of these regions along with their dependency over agriculture. For this reason, pilot survey was necessary. Then, we decided the bids.

[6]Obviously to decide which houses are to be surveyed we followed random sampling.

case, when the result of the dice-throwing was either one of the three available odd numbers – 1,3,5 – the respondent was offered Rs. 15 which is the minimum amount of the two bids. Again, when the outcome of the second dice throw was one of the three possible even numbers – 2,4,6 – corresponding household or respondent was asked for the upper value of the available two bids which, in this case, was Rs. 25. On both occasions, respondent was offered only one bid.[7] When the household accepted any of the offered bids, then that was his offered value or willingness or eagerness to pay for the conservation cum use of agriculture. But if any respondent refused to accept the bid, then his or her willingness to pay (WTP) was considered as zero for the proposed change and that specific respondent was regarded as a 'protest bidder'. Despite zero offerings, they cannot be kept out of the analysis because they will also enjoy the benefits of the proposed change. Similar procedure was followed for other two groups as well, that is, for 'middle bidding group' and 'higher bidding group'.[8] Hence, a reply of 'YES' to an offered bid was considered as the WTP (maximum WTP) of that specific respondent and a reply of 'NO' to a concerned bid made the respondent a 'protest bidder' (Bateman et al., 2002).

For the purpose of evaluation of agriculture, we have surveyed 500 households and out of those 500 respondents, we have found 302 respondents as willing and 198 respondents as 'protest bidders'. These respondents offered 'zero WTP values'[9]. Here we have made use of random utility model for describing the background theory or workings of the bid process of this work which very much puts the choices of an individual or respondent in a market situation, though the proposed market in this case is a hypothetical one.[10]

Now, we move to the econometric specification of our DC model. Our intention is to get an idea or estimate the mean WTP for 'conservation cum sustainable' upgradation of agriculture of Indian Sundarbans. For this purpose, logit model has been used for estimating the values of the coefficients of the determinants of people's decision regarding acceptance or rejection of an offered bid. We know that the probability distribution of the error term helps in the choice of the model for this kind of analysis. We have used the logit model as it has been mostly used in the literature of DC model and also it is simple to understand. Since we are applying logit model, our dependent variable is binary where an 'YES' to an offered bid was assigned the value '1' and a 'NO' was assigned the value '0'. Hence, it can be written as

---

[7]Since we are using single-bounded DC model.

[8]Similar process of bidding game involving double-dice throw was followed for fishery as well, though the amounts of bids were different.

[9]For various reasons.

[10]This random utility model describes the theory or methodology that was followed by Harou, Markandya, Bellu and Cistulli (1998). One more important thing is this process of random utility model has been followed for conducting CVM analysis for both agriculture and fishery.

$$\ln \frac{p_i}{1-p_i}$$

or

$$\frac{\ln\{p(\text{yes})\}}{\ln\{p(1-\text{yes})\}}$$

The log-odds ratio is the ratio of chance or probability of accepting an offered bid, which also reflects the WTP (P_i) of the respondent and the probability of non-acceptance of a bid ($1 - P_i$) or the probability of non-willingness of accepting an offered bid. One can also see that as probability of accepting an assigned bid increases, this ratio also goes up. The independent variables of our study are described in Panel A of Table 3.1 followed by the results in Panel B. Majority of the independent variables of our study are considered following the existing works on contingent valuation where mainly socio-economic variables and attributes are considered along with the offered bid.

In Table 3.1, Panel B, we have discussed the results of the logit model followed by the marginal effects.

Table 3.1. Independent Variables Used for the Valuation of Agriculture with Description and Results of the Logit model for Agriculture of Indian Sundarbans.

A: Independent Variables used for the Valuation of Agriculture with Description	
DC bid	Bids vector of Rs. 15, Rs. 25, Rs. 40, Rs. 50, Rs. 65 and Rs. 75
Income	Total monthly income from all sources
Family size	Household size
Age	Age of the respondent
Edu years	Total years of education of the respondent
Sex	Dummy variable; 0 for males and 1 for females.
Caste	Dummy variable; 0 for General Caste, 1 for OBC, 2 for SC and 3 for ST
Land holdings	Dummy variable; 0 for having land and 1 for not having any.
Agricultural dependency	Dummy variable; 0 for dependent upon agriculture directly or indirectly, and 1 if the family is not dependent on agriculture.
Mechanization	Dummy variable; 0 if the family has access of mechanization of any form and 1 for no access of mechanization.

(*Continued*)

Table 3.1. (*Continued*)

B: Results of the Logit model for Agriculture of Indian Sundarbans.

Variable	**Coefficient**	**Marginal Effects (*dy/dx*)**
DC bid/single-bounded bid	−0.003248*** (−19.25)	−0.006231*** (−7.22)
Income	0.003536*** (8.23)	0.001235*** (5.24)
Family size	0.000398*** (7.23)	0.000214*** (8.26)
Age	0.002489 (1.33)	0.0002580 (0.58)
Edu years	0.002357*** (5.05)	0.000347*** (3.43)
Sex	0.00259 (1.32)	0.009875 (0.79)
Caste	−0.00459*** (−5.27)	−0.00075*** (−4.89)
Land holdings	0.005459*** (7.85)	0.000320** (2.07)
Agriculture dependency	0.02897*** (5.75)	0.00269*** (4.26)
Mechanization	0.02478** (2.02)	0.00259* (1.90)
Constant	−9.24150*** (−7.26)	
Log-likelihood	−278.3254	
LR chi-square	312.20	
Prob > chi-square	0.000	
Pseudo-R^2	0.7950	
Total no. of observations	500	

Source: Estimated by authors.
Values in the parentheses indicate *t*-values.
*Denotes significance at 10% levels.
**Denotes significance at 5% levels.
***Denotes significant at 1% levels.

From the table it is seen that DC bids hold a negative and significant relationship with WTP which is understandable; as the amount of bid rises, probability to say YES falls. Income has a positive as well as significant impact because an increase in income increases the probability to accept the offered bid. Family size has a positive relationship because the larger the family size, the higher is the agriculture dependency. Similar thing can be said about the direct or indirect agriculture dependency. When a family depends on the agriculture and its products by one way or the other, then their WTP certainly rises. Similarly, higher years of education create sustainable awareness, which is expected to encourage educated people for contributing towards the development of agriculture. Mechanization and land holdings have a positive relationship with the bids because of the fact that people having land depend on agriculture and they seek for further improvement. Caste has a negative sign that implies when a movement from people of ‘unreserved’ caste to that of ‘reserved’ caste occurs, their WTP falls, perhaps because of their poor financial condition. Age and sex are insignificant perhaps because dependency is across all ages and sexes. One can say our model as a well-fitted or good-fitted one, depending upon the value of R^2 which is as high as almost 80%. For a cross-section database, this value is really good.

As multiple socio-economic indicators have been used by us as our explanatory or independent variables, the right approach is to check the existence of multicollinearity among the variables. In our case, we have found that the mean variance inflationary factor (VIF)[11] is 2.45, which is within the limit of tolerance[12] because the rule of thumb says that if VIF lies between 1 and 5, then it is of moderate level and it can be accepted.

Next, the mean WTP for sustainable development of agriculture has been estimated by us for the close-ended dichotomous choice model of our study. The average WTP of our model for agriculture is Rs. 33.20 per month in which the lower bound is of Rs. 19.25 and the upper bound is of Rs. 41.32. For a poverty-stricken community, the willingness to bear the amount of Rs. 33.20 on a monthly basis for the sake of sustainable development of agriculture can be considered to be moderate enough.

In Table 3.1, Panel B, we have also measured the marginal effects that show how a change in the WTP of an individual occurs when there is a change in any of the regressors (X_i, $i = 1, 2, \ldots, n$), keeping the other regressors constant.

Now we focus upon open-ended referendum where respondents were offered open-ended bids in which they were directly given the opportunity to express their max WTP for the conservation cum use of agriculture. This technique has an advantage in the sense that it helps to bring out the actual or real WTP of the respondent. In this technique, open-ended bids are directly regressed against the independent variables to find out the significant determinants of WTP. Ordinary least square (OLS) method is used for the purpose of this regression. We have also used the same variables like that of logit analysis of close-ended referendum. Regression results by applying OLS method, following the open-ended referendum, are reported in Table 3.2.

Both forms of applications – close-ended and open-ended models – give us almost similar sort of outcomes. Significant variables of the close-ended logit model analysis are also significant in open-ended OLS analysis as well. The value of R-square of OLS analysis is also above 71% which is, again, very good for a cross-section data set.

Finally, from OLS analysis we have found that the maximum WTP is Rs. 37.87 per month, which is higher than that of close-ended referendum. We have taken an average of the two WTPs obtained from the two forms of analysis and named it as *true or real WTP*, which is Rs. 35.53 per month. For a financially backward area like Sundarbans, this willingness of incurring a cost of above Rs. 35 per month is really significant and worth a find for the policy-setters. It also shows how immensely people value the development of agriculture in these parts.

---

[11] VIF measures the degree of collinearity among the regressors in logit analysis.
[12] 1/VIF is known as the level of tolerance.

Table 3.2. Regression Results of Open-Ended Referendum for Agriculture of Indian Sundarbans.

Variables	Coefficients	*t*-Values
Constant	−3.3689***	−3.69
Income	0.06985***	10.25
Family size	0.02325***	4.15
Age	0.02489	1.02
Edu years	0.02654**	2.09
Sex	0.03599	0.80
Caste	−0.00358***	−4.31
Land holdings	0.03321***	−7.21
Agriculture dependency	0.0526***	14.25
Mechanization	0.03752**	2.07
Dependent variable: max WTP (open-ended maximum WTP)		
$N = 500$, $F = 141.33$, prob > $F = 0.000$, adjusted $R^2 = 0.7123$		

Source: Estimated by authors.

**Denotes significance at 5% levels.

***Denotes significant at 1% levels.

Valuation of Fishery: Methodology and Results

Now we shift our focus towards evaluation of fishery of Indian Sundarbans. Being an area of several islands connected by small rivers and canals, it is quite natural that the primary occupations for people of these areas are fishery – both inland cultivation as well as open access catching.

As we have already mentioned in Section 2 of this chapter, for sampling, we have followed 'stratified random sampling', covering two important fishery-dependent blocks – Basanti and Namkhana and our sample size for valuation of fishery resource is 500. The process of CVM and its application by means of using close-ended single-bounded DC model or single-bounded bid offerings through groupings of respondents according to amounts of bids are same as that applied for agriculture. Only difference with agriculture is that the bids are higher than that of agriculture, which we have done deliberately because during survey we have witnessed that earnings from fishing are higher than that from agriculture. All these facts along with the views of members of fishermen association and fishermen themselves were considered while fixing the bids. In case of fishery the bids are Rs. 30 and Rs. 40 (*for the lower bidding group*), Rs. 60 and Rs. 70 (*for the middle bidding group*) and Rs. 90 and Rs. 100 (*for the higher bidding group*).

Even in places such as Basanti and Namkhana where people are dependent upon both agriculture and fishery, we have found that people consider fishery as their primary occupation for its round-the-year availability and good income-generating probability.

In short, although people are aware of the sustainable development of agriculture and they are eager to do so as seen from Section 2, they are more dependent upon fishery. We could literally see the enthusiasm and encouragement among people regarding their willingness to contribute towards the conservation of fishery. This willingness may have come from the fact that fishery is much more revenue-generating than agriculture. There are agitations and severe complaints against 'bottom trawling' and 'illegal fishing' but at the same time, it expresses their sense of urgency and awareness about protecting this important resource for the sake of their present and future generations. The fact that several people are directly and indirectly dependent on fishery in the form of not only fishing job but also for boat, fishing-net-related business, as parts of the huge supply chain of the collected fish that reach distant parts of the state, the country has definitely created awareness and willingness among people for its protection.[13]

Here also we have followed the *dual dice throwing process* before a bid was offered to the respondents. Depending on the acceptance or rejection to accept the offered bid, the respondents were classified as willing or protest bidders and accordingly their max WTP or zero WTP were recorded. Out of 500 respondents we have surveyed for valuation of fishery, we have found 346 respondents were found *willing* and 154 respondents were *unwilling* or *protest bidders*. Again, we cannot neglect the responses of protest bidders. Hence, we have considered both a YES that is considered as max WTP of a respondent and we have also considered a NO referring to a zero WTP. The dependent variable, again, like the model of agriculture is the log-odds ratio. For clarity, we once again mention the dependent variable:

$$\ln \frac{p_i}{1-p_i}$$

or

$$\frac{\ln\{p(\text{yes})\}}{\ln\{p(1-\text{yes})\}}$$

Here we have used the same variables like that in the analysis of agriculture but we have included fishery dependency instead of agriculture dependency and land holdings are not considered in this analysis.[14] Rest of the variables with their definition are unchanged like that of Table 3.3, Panel A. By fishery dependency

---

[13]Though this sense of awareness for sustainable use of this resource is absent in all fishermen communities this leads to various forms of conflicts and problems.

[14]Fishery dependency, which has been used here as a dummy variable, is defined as 0 for families having dependency upon fishery, either direct or indirect, and 1 if the family is not dependent on fishery.

Table 3.3. Results of the Logit Model and Open-Ended Referendum for Fishery of Indian Sundarbans.

A: Results of the Logit Model for Fishery of Indian Sundarbans

Variable	**Coefficient**	**Marginal Effects (*dy*/*dx*)**
DC bid/single-bounded bid	−0.005547*** (−7.24)	−0.002442*** (−5.23)
Income	0.002413*** (4.01)	0.000147*** (3.89)
Family size	0.001789* (1.91)	0.000327* (1.79)
Age	0.007025** (2.09)	0.000140* (1.79)
Edu years	0.003201*** (11.23)	0.000301*** (5.21
Sex	0.001021 (0.54)	0.000200 (0.34)
Caste	−0.004261 (−1.02)	−0.000367 (−0.99)
Fishery dependency	0.023346*** (11.25)	0.008014*** (7.23)
Constant	−5.201490*** (−7.25)	
Log-likelihood	−278.3254	
LR chi-square	310.57	
Prob > chi-square	0.000	
Pseudo-R^2	0.7254	
Total no. of observations	500	

B: Regression Results of Open-Ended Referendum for Fishery of Indian Sundarbans

Variables	**Coefficients**	***t* values**
Constant	−4.2136**	−2.03
Income	0.02034***	5.23
Family size	0.0903***	3.24
Age	0.0410**	1.93
Edu years	0.0033***	4.23
Sex	0.0209	1.01
Caste	−0.0683	−0.62
Fishery dependency	0.0342***	11.26

Dependent variable: max WTP (open-ended maximum WTP)

N = 500, F = 133.68, Prob > F = 0.000, adjusted R^2 = 0.74.26

Source: Estimated by authors.

Values in the parentheses indicate *t*-values.

*Denotes significance at 10% levels.

**Denotes significance at 5% levels.

***Denotes significant at 1% levels.

we mean not only association in the form of fishermen or owner of fishing boats or inland fisheries but also as a person who is anyway connected with either or both forms of fishery (*inland and open access*) such as fishing net, boat repairing, pipe for water supply and even as a part of the supply system of fishes.

The outcome of the logit analysis of close-ended DC model is reported in Table 3.3, Panel A, followed by the results of the open-ended regression analysis in Table 3.3. Panel B.

The results of our logit model along with the effects of marginal changes are reported in Table 3.3, Panel A. It gives us very much expected results. Income, family size, age, education and fishery dependency – these variables bear positive as well as significant impacts on the WTP. These results are very much expected because along with increase in income, people are expected to contribute towards the conservation of fishery and hence their willingness to accept an offered bid is supposed to be high. Similarly, larger family is supposed to have more earning members and fishery being the prime occupation in the area is expected to generate employment opportunities to the family members. On several cases, we have seen that multiple members of a single family are fishery-dependent to earn their livelihoods. Hence, larger family members are expected to have higher WTP. Age and education are significant that implies the fact that along with age and education, people's awareness on sustainable fishing and its conservation grows. Likewise, dependency on fishery, directly or indirectly, certainly increases the probability to accept an offered bid.

Similarly, the negative sign carrying significant variables also has meaningful implications. An increase in the value of bid is certainly going to have a negative impact on the WTP which is seen here. The fact that caste and sex are insignificant signify the general dependency of people on fishery – across all castes and sexes. Similar sort of outcome is observed in case of marginal effects as well.

The value of pseudo-R^2 of our logit analysis is above 72.5%, which can be considered to be very good for a model based on cross-section data base.

Next, we have tried to estimate the mean WTP for our logit model of fishery. The findings are that the average WTP for sustainable protection cum development of both inland and open-access fishery of Sundarbans is as high as Rs. 86.46 on a monthly basis, with the lower bound of Rs. 72.98 and upper bound of Rs. 91.29. Given the financial backdrop of the region we have surveyed, this value can be considered to be really high and good which reflects the eagerness of the people of Sundarbans to save their prime natural source of livelihood.

Once again, we have tested for the existence of possible multicollinearity in our model. The mean VIF of the dependent variables is 3.21 which again is within the permissible limit for being below 5. Next, we have looked to estimate the determinants of WTP by applying open-ended referendum in panel 3B. Hence, like the agriculture model, once again, we have asked the respondents to state their max WTP for conservation of fishery by open-ended referendum. The outcomes are obtained by regressing the values of WTP on the regressors by applying OLS. The results are reported in Table 3.3, Panel B.

Here also both in logit model and in open-ended regression model, similar variables are significant though their level of significance may vary. One interesting fact is that in open-ended referendum, the value of adjusted R^2 is as high as over 74% which is very meaningful for our analysis. Also, the average WTP from regression analysis by application of open-ended referendum is Rs. 92.25 monthly which is meaningfully higher than that of closed-ended referendum. We have also taken an average of the two mean WTPs to find out the *Actual or True WTP* for fishery, which is Rs. 89.35, which is highly meaningful because in an area like Sundarbans, people are willing to contribute almost Rs. 90 on a monthly basis for the sustainable use and growth of fishery deserves acknowledgement from the policy-setting bodies.

Concluding Remarks

Our emphasis has been on the socio-economic aspects of people of Sundarbans and in this study an attempt has been made to relate these aspects to the valuation of two most commonly used as well as popular natural resources on which people of Sundarbans largely depend on agriculture and fishery for survival. We have applied the very popular and appropriate methodology for valuation of natural resources CVM to get an idea about the valuation of these two resources by the stakeholders themselves (Chatterjee, 2022).

In our analysis, it has been witnessed that people of Sundarbans are not only steadily reliant upon use of these two resources for livelihood generation but they are very much eager to protect and sustainably use these resources as well, obviously except the problem of 'illegal bottom trawling' that cannot be neglected and needs to be highlighted to the policy-setters for the sake of future generation and survival of open access fishing. But overall people are very much enthusiastic to conserve and sustainably use these resources as reflected from their high values of WTP.

People in that area understand the interlinkages between agriculture and fisheries and therefore their eagerness can be seen to protect both the resources by switching the uses of lands, according to the need of the season. Thus, they ensure optimal uses of land between fishery and agriculture.

The Government and policy-setters, therefore, have ample scopes to use this study as reference and take necessary, suitable action. The government should focus on that and work accordingly to conserve agriculture and ensure its proper development by means of irrigation; mechanization and non-saline elements. Regarding fishery, one can say that stopping of 'illegal bottom trawling' should be the first priority of the regulating agencies. Regarding inland fishery, we need more awareness, more trainings to combat the challenges of global warming and climate change. People of Sundarbans are both aware as well as eager to conserve fisheries of both forms – they need proper support from the regulating agencies and governments.

Lastly, high values of WTPs ensure the fact that if governments and NGOs work efficiently enough, together, for the conservation cum sustainable development of these resources then it could be possible to achieve the Sustainable Development Goals in these parts of the 'UNESCO Heritage' Sundarbans.

References

Arrow, K., Solow, R., Portney, P. R., Leamer, E. E., Radner, R., & Schuman, H. (1993). Report of the NOAA panel on contingent valuation. *Federal Register*, *58*(10), 4601–4614.

Bateman, I., Carson, R. T., Day, B., Hanemann, W. M., Hanley, N., Hett, T., Jones, A., Loomes, G., Mourato, S., Ozdemiroglu, E., Pearce, D. W., Sugden, R., & Swanson, J. (2002). *Economic valuation with stated preference techniques: A manual*. Edward Elgar.

Chatterjee, N. (2019). Use of renewable resources and sustainable livelihood in Indian Sundarbans: An attempt through contingent valuation method. *Vidyasagar University Journal of Economics*, XXII, 98–117.

Chatterjee, N. (2022). 'Households' willingness to pay for improved water supply system in the dryland areas of West Bengal – An estimation using double-bounded dichotomous-choice model. *Economic Affairs*, *67*(5), 787–795.

Chatterjee, N. (2023). Climate change and labour force migration in Indian Sundarbans: An econometric exploration. *Inclusive*, *II*(24), 372–386.

Gupta, K. (2005). Mangrove-fishery linkages under alternative policy regimes. In N. Sengupta & J. Bandyopadhyay (Eds.), *Biodiversity and quality of life* (pp. 272–293). Milan India Ltd.

Haraou, P., Markandya, A., Bellu, L., & Cistulli, V. (1998). *Environmental economics and environmental policy: A workbook*. Economic Development Institute (EDI) of World Bank.

Singh, V., & Roxy, M. (2022). A review of ocean-atmosphere interactions during tropical cyclones in the north Indian Ocean. *Earth-Science Reviews*, *226*, https://doi.org/10.1016/j.earscirev.2022.103967

Chapter 4

An Analysis of the Price Behaviour of Agricultural Commodities in India

Debraj Das and Debabrata Mukhopadhyay

Department of Economics, West Bengal State University, India

Abstract

Volatility in agricultural commodity prices assumes a lot of significance since its associated uncertainty is one of the major factors affecting the income security of producers and traders. The study aims to investigate the fluctuations and instability of the two crops, paddy and wheat, in the states of India. It has estimated instability in prices over the years (2006–2023) for paddy and wheat for the top three producing states by using coefficient of variation (CV), Cuddy-Della Valle index and Coppock's index. Then it estimates volatility by generalized autoregressive conditional heteroskedasticity (GARCH) framework. It is observed that the average standard deviation of prices across mandis for paddy in a given month is 0.18 and for wheat it is 0.22. Both for paddy and wheat Punjab has the maximum instability among the three states if we consider CV and Coppock's index. But if we consider the Cuddy-Della Valle index, Madhya Pradesh has the maximum instability for wheat. The noticed unpredictability in the price series of paddy revealed that the value of ARCH (α) has significant term and value of GARCH (β) was also significant for all the price series. The volatility ($\alpha + \beta$) was quite persistent of the order of 0.985 in case of Uttar Pradesh. In case of wheat Punjab has the maximum value 0.95 signifying explosiveness. In case of wheat Punjab has the maximum value 0.95 signifying explosiveness.

Keywords: Agricultural price; AGMARKNET portal; intrastate variation; volatility; Cuddy Della Valle index; Coppock's instability index; ARCH; GARCH

Sustainable Agricultural Practices: Economic and Environmental Implications, 43–55

doi:10.1108/978-1-83608-336-820241004

Introduction

Fluctuations in agricultural price become a common problem to the countries. But it may not cause any serious problem if they follow a specific trend like a seasonal pattern. But in Indian context fluctuations in agricultural commodities cause a serious hindrance to all the sections of people involving the farmers, traders and also the consumers. Large fluctuations in agricultural prices are particularly painful towards the economically backward section of the country. These kinds of fluctuations cause a tremendous disorientation in their daily life. Highly volatile agricultural prices become very problematic for the government to formulate a specific policy to improve the well-being of the farmers. The problem aggravates in case of a developing country like India where majority of the work force still depends on agricultural activities. Uncertainty in receiving a proper return against their toil the farmers' morbidity level becomes very high and sometimes it leads to farmers' suicide. Hence measuring the extent and nature of volatility in agricultural prices is of utmost importance.

Theoretically, volatility refers to an uncertain movement of a random variable over a period of time. Volatility in agricultural commodity prices assumes a lot of significance since its associated uncertainty is one of the major factors affecting the income security of producers and traders threatening the performance of agriculture and welfare of the consumers. Hence, volatile agricultural commodity prices in general and food price in particular are a policy priority engaging the attention of economists and policy-makers.

Commodities have been guiding and determining the fate and fortune of nations because of the volatility prevailing in their markets. It has been argued widely that inter alia, volatility in agricultural commodity prices originates from the population and income-induced demand increase following years of good economic growth accompanied by changing food habits but not accompanied by commensurate supply, futures trade and globalization. However, volatility in agricultural commodity prices is mainly attributed to supply shocks coupled with the short-run demand and supply elasticity coefficients. Futures market reveals that information on prices, hedging and speculation, and physical transaction influence the volatility in prices. But increasing price volatility has made speculation common and questioned the utility of futures trading in agricultural commodities. In an arbitrage-free economy, asset price in market will rise due to increasing flow of information, thereby generating the volatility. The WTO-led liberalization exposed agriculture to global competition resulting in international price volatility being transmitted to domestic prices (Goleti & Babu, 1994). The above argument merits a detailed empirical analysis in India's story of food price volatility.

Literature Review

Several studies have investigated about the instability in agricultural prices in India and outside India. Chand and Raju (2008) have studied volatility in the agriculture sector of Andhra Pradesh into different component parts. In case of a large state like Andhra Pradesh, the instability gathered from state-level data differs significantly from those at the disaggregated level. Thus, the state-level analysis fails to give the whole image of shocks in agricultural production, which misjudges shocks in farm income.

Mahir and Abdelaziz (2011) had conducted a study to measure the extent of instability and contribution of different components to change in mean production of the main crops grown in the Gezira scheme. The study included four agricultural commodities like sorghum, wheat, cotton and groundnuts. The study separated the time series data into two parts. The first period consists of the years before the adoption of liberalization policy (1970/1971 to 1991/1992) and the second period consists of the years after the adoption of liberalization policy (1992/1993 to 2007/2008). Sorghum, wheat and cotton had observed a steady increase in instability over the two periods and there was decline in instability of groundnut during post-liberalization period. It is seen that instability in area and yield of all the crops pursued a similar bearing. Recovery of water system channels, solid horticultural exploration and augmentation were recommended to assist with accomplishing steadiness in agricultural production. Kumar and Jain (2013) have looked through the patterns in growth and instability in Indian agriculture at the local level. Enormous dispersions were seen in the productivity of crop sector inside the state and for the country overall. The data demonstrated the important role of inputs in improving the productivity of crop sectors such as rainfall, fertilizers, human resource, irrigation and transportation. Pichad et al. (2014) have examined the growth and instability in production, area and productivity of chickpea in the context of Amravati district. Compound annual growth rate as well as coefficient of variation was utilized for the analysis. From the results it was revealed that compound growth rates for area, production and productivity were found positive and significant. The coefficient of variation (CV) revealed that instability in the area for chickpea showed less variation than production and productivity at entire period whereas production witnessed highest instability if compared with area and productivity at overall period. Ayalew and Sekar (2016) have calculated the patterns and territorial disparity of maize production in India. Cuddy Della Valle index, compound annual growth rate and decomposition analysis were used. The data range was from 1980 to 1981 and from 2011 to 2012. The study found an increase in the area and production of maize during the period. Such increment was possible simply because of expansion in yield. The study established that maize performed better in Bihar, Andhra Pradesh, Gujarat, Rajasthan and Uttar Pradesh as expansion in yield is combined with decrease in instability because of the adoption of modern varieties of maize. Tewari et al. (2017) have calculated the instability and growth in yield, area and production of wheat in Uttar Pradesh. Decomposition analysis and Cuddy Della Valle index were used for the period 1990–1991 to 2013–2014. The maximum productivity was tracked down in western area and lowest in Bundelkhand region. Also, there was soaring instability in yield and production when contrasted with area under wheat and yield effect was found to be the central force in the growth of production. Ismail et al. (2017) analysed the price volatility of agricultural commodities and food in Pakistan's context. They investigate the factors responsible for the volatility of agricultural commodity prices. They used monthly data from April 1983 to April 2013 for their analysis. generalized autoregressive conditional heteroskedasticity (GARCH), EGARCH and GJR models were estimated for all the variables using normal and Student's *t*-distribution. The results concluded that the mean and the volatility effect of interest rate and exchange rate were transmitted across all the selected commodities. Volatility observed in the price of fertilizer was only transmitted on the volatility of sunflower oil. Later, analysis showed that past price had significant impact on current prices for all the commodities except for cotton, soybean oil and sunflower oil. Huchet-Bourdon (2011) in his OECD report tried to find out the question of whether commodity price volatility has considerably changed with the rapid run up in world prices in 2006–2009, followed by an equally sharp decrease in many commodity prices. The report analysed international price volatility for some agricultural commodities over the past half-century and their association with crude oil, fertilizer and the euro-dollar exchange rates. The inquiry utilized different data sources, frequency of price observations, periods of observation, price volatility measures and a number of statistical tests to examine the various magnitude of the issue. Bathla (2012) in her work discussed how trade liberalization and globalization had uncovered agricultural sector of many developing countries to unexpected disturbances, caused not simply by demand-supply conditions inside their economies but also by volatility in global commodity prices, exchange rate and flow of imports. The work assessed the extent of sensitivity of Indian agriculture to these elements and investigates strategy choices that may nullify their negative impacts and maintain price stability and incentives. The analysis is done for one important tradable product viz. wheat by using a structural econometric model, under the exportable and importable conditions from 1980–1981 to 2009–2010. Discoveries uncover wheat to be progressively determined by an impetus structure in view of its association with exchange rate, world price and other factors. Counterfactual reproduction tests show that because of trade and sector-specific policies, wheat price and output tend to be more versatile to fluctuations in international price and other shocks contrasted to its exports and imports. Burman et al. (2018) showed that agricultural markets in India have suffered from high price volatility. There may be an aspect of a Samuelson Cobweb Model at work, which generates a cycle of boom and bust. When food prices are high, consumers protest, and in the years, when food prices are low, farmers have suffered and demanded loan waivers. Four policy pathways deal with the cobweb model: storage, national trade, international trade and futures trading. We disagree that the Constitution imposes an obligation upon the Union government to achieve a national market. They find out an implementable set of steps through which the Union government can attain a national market for agricultural produce. Kalkuhl et al. (2016) in their book illustrated the stability dimension of food security and in particular, the causes, consequences and remedies related to extreme events in food markets. It argued that price volatility is an intrinsically market-related economic concept. Different articles of this book carefully embedded into the political, agricultural, climate and nutritional domains. Sekhar (2003) believed that price volatility of agricultural commodities assumes critical importance with regards to the continuous discussion regarding agricultural trade liberalization in India. The contentions against agricultural trade liberalization were in many cases based on the issue of bigger volatility in international markets. In order to make informed judgments about this crucial aspect of agriculture, which had implications for the entire economy, it was essential to study the volatility patterns in international and domestic markets in a comparative framework. The study, using monthly price data, finds little evidence to show that the international agricultural prices were uniformly more variable than the domestic prices. The study showed that the intra-year variability was higher in domestic markets while the inter-year variability was higher in the international markets. The current bound rates of duty were generally found adequate in this study except in case of soybean oil and sugar.

Kumar and Sharma (2006) assessed government price policy by using monthly indices of wholesale prices of wheat, rice and coarse-grains. They had analysed the yearly price data which depicted that in the 1990s the inter-year variability in annual nominal prices declined both for wheat and rice when compared with the data set of 1980s. The changeability declined for the real price of rice; additionally, the real price of wheat and nominal and real price of coarse grains showed expanded fluctuation during the 1990s in contrast with 1980s. Examination of month-to-month prices uncovered that intra-year changeability shot up for wheat while it descended for rice during the 1990s in contrast with 1980s. The ascent in cost wedge in wheat and in course cereals during the 1990s was brought about by ascends in occasionally, pattern and repetitive sporadic variables. Because of rice, then again, the commitment of these three elements was not stable. The relapse results showed that the public authority activities of trading were more fruitful in diminishing the occasional reach in wholesale prices on account of rice when contrasted with wheat. Varieties in production of coarse grains likewise fundamentally caused occasional climb in price of wheat and rice; however, they were more close substitutes to wheat than to rice. Any deviation of cereals' accessibility from long haul pattern prompted higher government mediation regarding acquirement and payment of wheat and rice. Likewise, deviations in own result of wheat and rice additionally prompted higher mediation by the public authority in these two items. Deviations in precipitation, yield and wholesale prices were the huge factors, which loaned instability into the production of wheat and rice. Kumar and Jha (2017) inspected the transmission system of an increment in energy prices on the price of chosen agricultural commodities in India utilizing monthly wholesale price indices during from April 1994 to March 2014. In order to assess the effect of deregulation of some petroleum products which was implemented since April 2002, the sample data was divided into two periods (April 1994–March 2004 and April 2004–March 2014), besides analysing for full period so that before and after period analysis will provide clearer picture of possible link between the two prices. To supplement the finding of Johansen co-integration analysis, we assessed the nature and extent of causal relationship between the variables. The results shown proof of equal development between price of energy and all chose agricultural products after liberation, and that implies higher transmission between unrefined petroleum and commodity prices. The outcomes acquired are supposed to help in comprehension of transmission system for strategy creators to streamline and balance out the markets. Bhoi et al. (2019) analysed the data collected through a pan India primary survey of farmers, traders and retailers to examine the role of mark-ups charged by different intermediaries in the agricultural markets in creating a wedge between farm gate prices and retail prices for 16 major food items covered in the Consumer Price Index-Combined (CPI-C) basket. The average share of farmers in the consumers' rupee is found to be in a range of 28 and 78 per cent for different food items. Empirical results suggest that factors contributing to greater efficiency in the supply chain such as better road network, mandi infrastructure, tele-density to improve flow of information, irrigation facilities to reduce supply uncertainties, and increase in overall literacy levels in the country enabling greater consumer awareness can help reduce mark-ups. Varshney et al. (2020) have assessed the impact of the spread of COVID-19 and the lockdown on wholesale prices and quantities traded in agricultural markets. We compare whether these impacts differ across non-perishable (wheat) and perishable commodities (tomato and onion), and the extent to which any adverse impacts are mitigated by the adoption of a greater number of agricultural market reform measures. They use a granular data set comprising daily observations for three months from nearly 1,000 markets across five states and use a double- and triple-difference estimation strategy. Expectedly, our results differ by type of commodity and period of analysis. While all prices spiked initially in April, they recovered relatively quickly, underscoring the importance of time duration for analysis. Wheat prices were anchored in large part by the minimum support price, while tomato prices were lower in some months. Supply constraints began easing in May with greater market arrivals perhaps reflecting distress sales. Market reform measures did help in insulating farmers from lower prices, but these effects are salient for the perishable goods, and not so much for wheat where the government remained the dominant market player. Taken together, these results point to considerable resilience in agricultural markets in dealing with the COVID-19 shock, buffered by adequate policy support. Anoopkumar (2014) narrated that the commercial agriculture sector of India is the heightened instability in price both at the inter-year and intra-year level. An exploration of intra-year price instability, which is rarely addressed in the empirical studies, is attempted in this work by selecting the case of four major commercial crops of India namely, small cardamom, black pepper, natural rubber and tea. The study concludes that even in the scenario of integrated market setting price instability at the intra-year level is guided mainly by the seasonal character of supply. Hence, market offers high price to the farmers only when they have less to offer in the market and vice versa, which makes their income unstable in the short run, highlighting the need for stabilizing producers' returns.

Dudhat et al.'s (2017) study was carried out by using time series data of price from 1996 to 2016, to compare the price variation, to measure the price instability and price seasonality of major domestic groundnut markets of India i.e. Kurnool (Andhra Pradesh), Rajkot (Gujarat) and Villupuram (Tamil Nadu). Data were collected from Directorate of Economics and Statistics, Ministry of Agriculture, Govt. of India. The price variation was compared by using descriptive statistics. Range and standard deviation indicated wide variation in the price level in all the markets. The price instability and price seasonality based on trend model were measured. Estimates of fitted quadratic trend model for Kurnool and Villupuram markets and estimates of fitted linear trend model for Rajkot market were significant. Highest instability was observed for Rajkot market. Seasonal factors were observed highest in August, April and June for Kurnool and Rajkot markets, respectively. The farmers and policy makers can use this price signal for their future plan. Ghoshray and Ghosh (2011) in their article have examined the relationship between the various prices of wheat quoted at different market centres in four Indian states. They find evidence of M-TAR asymmetric adjustments of wheat prices, indicating that price signals within states are transmitted over time in an asymmetric manner. This type of price adjusting behaviour is consistent as to how price differentials may respond to poor dissemination of knowledge regarding market conditions and high transactions costs. Despite this huge literature the nature of instability in agricultural commodity prices using the Mandi data is still an unexplored area. In this context we try to carry out our analysis.

Data and Methodology

Data

Secondary data has been collected from AGMARKNET Portal under Ministry of Agriculture and Farmers Welfare, Economic Survey (2021-22), different reports of RBI.

Estimation of Instability Index

The magnitude of instability in the prices was measured in relative terms by Cuddy-Della Valle index (Cuddy & Della Valle, 1978) which is used as a measure of variability in time-series data. Simple CV overestimates the level of instability in the time series data characterized by the long-term trends, whereas Cuddy-Della Valle Index corrects this:

$$\text{Cuddy-Della Valle index} = \text{CV} \times \sqrt{\left(1-\overline{R}^2\right)}$$

where CV is the coefficient of variation in percentage and regression adjusted by the number of degrees of freedom and $\overline{R}^2$ is the coefficient of determination from a time trend.

Coppock's Instability Index

The CV represents the ratio of the standard deviation to the mean, and it is a useful statistic for comparing the degree of variation from one data series to another, even if the means are drastically different from each other, it is expressed as percentage.

The simple CV often contains the trend component and thus over estimates the level of instability in time series data characterized by long-term trends. To overcome this problem, the instability index given by Coppock's instability index of variation is used. Coppock instability index is a close approximation of the average year-to-year percentage variation adjusted for trend and the advantage is that it measures the instability in relation to the trend in prices. A higher numerical value for the index represents greater instability. The algebraic form of equation is

$$\text{CII} = [(\text{antilog}\sqrt{\log v} - 1) \times 100]$$

$$\text{Log}\, v = \sum \left[\frac{\log \dfrac{X_{t+1}}{X_t}}{N-1} - \frac{m}{N-1} \right]^2$$

where X_t is the area/production/yield in the year t, N the number of years, CII the Coppock's instability index, m the arithmetic mean of difference between the logs of X_{t+1}, X_t, etc., log v the logarithmic variance of the series.

Testing of Stationarity

The unit root or non-stationary test described by Dickey and Fuller (1979) is valid, if the time series y_t is well characterized by an autoregressive (AR(1)) with white noise errors. Many financial time series, however, have a more complicated dynamic structure than is captured by a simple AR(1) model. Said and Dickey (1984) augment the basic AR unit root test to accommodate general autoregressive moving average (ARMA(p, q)) models with unknown orders and their test is referred to as the augmented Dickey– Fuller (ADF) test. The ADF tests the null hypothesis that a time series y_t is $I(1)$ against the alternative that it is $I(0)$, assuming that the dynamics in the data have an ARMA structure. The ADF test is based on estimating the test regression:

$$\Delta y_t = \beta_1 + \beta_2 t + \delta Y_{t-1} + \Sigma \alpha_i \Delta Y_{t-i} + \varepsilon_t$$

where ε_t is a pure white noise error term and where $\Delta Y_{t-1} = (Y_{t-1} - Y_{t-2})$, $\Delta Y_{t-2} = (Y_{t-2} - Y_{t-3})$, etc. The number of lagged difference terms to include is often determined empirically. Lagged difference terms are included to make the make the error term serially uncorrelated. We test whether $\delta = 0$. This test was applied to the monthly price series of paddy (common) and wheat to test the null hypothesis that the series has unit root problem or the price series is non-stationary in nature.

Price Volatility by GARCH Model

We are also interested in volatility clustering in agricultural prices over time. This term indicates those periods in which their prices show wide swings for a longer period followed by periods in which there is relative calm. Major variations in prices lead to incorrect decision-making.

To measure the degree of price volatility, GARCH 1,1 mode following Bollerslev (1986) was used which can be specified as

$$\sigma^2_t = \alpha + \beta u^2_{t-1} + \gamma\sigma^2_{t-1}$$

Which says that the conditional variance of *u* at time *t* depends not only on the squared error term in the previous time period but also on its conditional variance in the previous time period.

The squared error follows a heteroskedastic ARMA (1,1) process. The autoregressive root which guides the persistence of volatility shocks in the price series is the sum of $\alpha + \beta$. The ARCH parameters correspond to α and GARCH parameters to β. If the sum of ARCH and GARCH coefficients close to 1, then it signifies that volatility shocks are fairly persistent.

Results and Discussion

We have started our analysis by looking upon the intrastate variation for two principal cereal crops of India: paddy (common) and wheat (see Table 4.1). For this analysis we focus on average monthly price data for wheat and paddy sold in any mandi in India between 2006 and 2023. It should be noted that the prices we analyse are

Table 4.1. Intrastate Variation in Real Prices for the States in 2023 for Two Principal Cereal Crops Paddy (Common) and Wheat.

State	S.D. (Paddy)	S.D. (Wheat)
Andhra Pradesh	0.16	0.23
Chhattisgarh	0.13	0.12
Gujarat	0.14	0.19
Haryana	0.21	0.21
Karnataka	0.18	0.24
Kerala	0.14	0.14
Madhya Pradesh	0.22	0.22
Maharashtra	0.16	0.15
Odisha	0.38	0.32
Punjab	0.26	0.42
Rajasthan	0.14	0.11
Tamil Nadu	0.21	0.23
Uttar Pradesh	0.11	0.14
West Bengal	0.15	0.19

Source: Authors.

wholesale prices observed at APMC mandis. It is very likely that these are not prices received by farmers, especially since it is large farmers who sell in mandis, small farmers are more likely to sell to local village intermediaries. However, our price data has several advantages. They are actual prices recorded at a high frequency and at a crucial stage in the supply chain – mandis are key points of aggregation. Other sources of price data are usually recalled estimates of unit values, geographically aggregated and very low frequency. We use log real prices [with the CPIAL (food) as the deflator] so that the variance of log prices is unit independent and makes it compatible for across country comparisons. The average standard deviation of log (real) prices across mandis for paddy in a given month is 0.18 and for wheat it is 0.22. For comparative purposes this is higher than Philippines: where for rice and corn (the main food commodities grown there), Allen (2014) found the standard deviation to be 0.15 in Philippines, which a country formed by group of islands and has high transport and information costs. Our variance estimate is robust to include all cereals. In case of paddy Odisha has maximum deviation whereas in case of wheat Punjab has maximum deviation. The variance in prices has been high since 2005 and hence the results are not capturing the effect of an outlier year. Moreover, we do not observe any trend in time-series of the standard deviation which implies that an increase in cell-phone penetration during this period does not appear to have had a causal effect on price variation in grains across India.

After finding out the intrastate variation we proceed towards the instability in prices over a span of 18 years (2006–2023) for the top three producing states of paddy and wheat. We have chosen 2006 as the starting year; from this year the AGMARKNET portal is fully functional and we can collect our required data for this period. We have collected monthly prices from different markets from AGMARKNET portal. Then we have checked out for the level of instability for the states. For finding out the instability we have chosen three indicators: CV, Cuddy-Della Valle index and Coppock's index (Table 4.2). We have used three indicators to ensure that no bias is present in our analysis. We can see very clearly that the results are quite consistent in nature. In case of paddy we can see that Punjab has the maximum instability among the three states. In case of wheat the situation is slightly different. Again Punjab has the maximum level of instability in case of CV and Coppock's index but if we consider the Cuddy-Della Valle index Madhya Pradesh has the maximum instability.

ADF test was used to determine the number of times the series needs to be differenced to make it stationary for the next round of analysis.

The analysis revealed that for both rice and wheat the price series from 2006 to 2023 were non-stationary in nature which means that they had unit root problem. But after first differencing they became stationary and then we applied the GARCH model. The values of the univariate GARCH (1,1) model were represented in Table 4.2 for all the commodity price series. The extent of volatility clustering was depicted through the ARCH and GARCH parameters, i.e., $(\alpha + \beta)$ of price series. Both the ARCH(α) and GARCH(β) parameters were found to be significant for paddy and wheat. The volatility $(\alpha + \beta)$ was quite relentless of the order of 0.985 in the case of Uttar Pradesh. For the other two states it is modest in nature. In case of wheat Punjab has the maximum value 0.95 of $(\alpha + \beta)$ signifying explosiveness.

Table 4.2. Instability and Volatility in Prices over the Years (2006–2023) in States.

	Paddy (Common)				Wheat		
			Instability in Prices				
State	**CV**	**Cuddy-Della Valle Index**	**Coppock's Index**	**State**	**CV**	**Cuddy-Della Valle Index**	**Coppock's Index**
West Bengal	21.98	21.65	32.65	Uttar Pradesh	96.63	31.88	52.32
Uttar Pradesh	20.86	18.38	36.84	Madhya Pradesh	17.07	58.34	35.65
Punjab	36.11	32.09	41.25	Punjab	110.95	11.38	65.23
			Estimate Volatility by GARCH Framework				
State		α	β	**State**		A	β
West Bengal		0.0416	0.6642	Uttar Pradesh		0.1004	0.1199
Uttar Pradesh		0.1442	0.7791	Madhya Pradesh		0.031	0.2115
Punjab		0.1591	0.5275	Punjab		0.7159	0.2335

Source: Authors.

Conclusion

From different parameters of instability, it is clearly visible that food price fluctuations were quite large during the period of 2006–2023. There were significant evidence of volatility clustering as well. The ARCH and GARCH parameters were not only significant but they also showed a particular trend as the GARCH parameters were higher than the ARCH parameters represented long persistence of volatility. The fluctuation existed not only in case of time series data but also in case of cross sectional intra-state analysis also both for paddy and wheat. This signifies that the overall trend regarding the fluctuation in the agricultural commodity prices is alarming in nature. Excessive fluctuation in the price signifies the continuous shocks the market is experiencing. These shocks are affecting the market signalling system to run in an efficient and smooth manner. Although government has taken several measures like the e-NAM scheme to enhance transparency in transactions, price discovery and farmers' reach to larger number of markets to sell their product to buyers of their choice are at their convenience. It was done to ensure the smooth functioning of the market of agricultural commodities, and it is not sufficient by any means. Addressing supply-side bottlenecks, particularly smoothing production, is important. India's price policy also needs to be carefully calibrated. Sustained increases in the price of cereals call for a more judicious procurement policy and timely liquidation of stocks.

References

Anoopkumar, M. (2014). Intra-year price instability of commercial crops in India: Exploring the underlying dynamism. *International Journal of Food and Agricultural Economics (IJFAEC)*, *2*(1), 145–156.

Ayalew, B., & Sekar, I. (2016). Trends and regional disparity of maize production in India. *Journal of Development and Agricultural Economics*, *8*(9), 193–199.

Bathla, S. (2012), Volatility in agriculture commodity prices in India: Impact and macroeconomic and sector-specific policy responses. In *123rd EAAE seminar*, Dublin.

Bhoi, B. B., Kundu, S., Kishore, V., & Suganthi, D. (2019). Supply chain dynamics and food inflation in India. *RBI Bulletin*, *73*(10), 95–108.

Bollerslev, T. (1986). Generalized autoregressive conditional heteroscedastic process. *J. Econometrics*, *31*, 307–327.

Burman, A., Roy, S., Shah, A., & Patnaik, I. (2018). Diagnosing and overcoming sustained food price volatility: Enabling a national market for food. *NIPFP Working Paper Series*, No. 236.

Chand, R., & Raju, S. S. (2008). Instability in Andhra Pradesh agriculture – A disaggregate analysis. *Agricultural Economics Research Review*, *21*(2), 283–288.

Cuddy, J. D., & Valle, P. D. (1978). Measuring the instability of time series data. *Oxford Bulletin of Economics and Statistics*, *40*(1), 79–85.

Dickey, D. A., & Fuller, W. A. (1979). Distribution of the estimators for autoregressive time series with a unit root. *Journal of the American Statistical Association*, *74*, 427–431.

Dudhat, A. S., Yadav, P., & Venujayakanth, B. (2017). A statistical analysis on instability and seasonal component in the price series of major domestic groundnut markets in India. *International Journal of Current Microbiology and Applied Sciences*, *6*(11), 815–823.

Ghoshray, A., & Ghosh, M. (2011). How integrated is the Indian wheat market? *Journal of Development Studies*, *47*(10), 1574–1594.

Goletti, F., & Babu, S. (1994). Market liberalization and integration of maize markets in Malawi. *Journal of Agricultural Economics*, *11*(1), 311–324.

Huchet-Bourdon, M. (2011). Agricultural commodity price volatility: An overview. OECD Food, *Agriculture and Fisheries Papers*, No. 52, OECD Publishing, Paris.

Ismail, A., Ihsan, H., Khan, S. A., & Jabeen, M. (2017). Price volatility of food and agricultural commodities: A case study of Pakistan. *Journal of Economic Cooperation & Development*, *38*(3), 77–120.

Kalkuhl, M., Von Braun, J., & Torero, M. (2016). *Food price volatility and its implications for food security and policy*. Springer Nature.

Kumar, A., & Jain, R. (2013). Growth and instability in agricultural productivity: A district level analysis. *Agricultural Economics Research Review*, *26*, 31–42.

Kumar, P., & Sharma, A. (2006). Price variability and its determinants: An analysis of major foodgrains in India. *Indian Economic Review*, *41*, 149–172.

Kumar, R. R., & Jha, G. K. (2017). Examining the co-movement between energy and agricultural commodity prices in India. *Journal of the Indian Society of Agricultural Statistics*, *71*(3), 241–252.

Mahir, M. E. A. E., & Abdelaziz, H. H. (2011). Analysis of agricultural production instability in the Gezira scheme. *Journal of the Saudi Society of Agricultural Sciences*, *10*(2), 53–58.

Pichad, S. P., Wagh, H. J., & Kadam, M. M. (2014). Growth in area, production and productivity of chickpea in Amravati district. *International Research Journal of Agricultural Economics and Statistics*, *5*(2), 289–292.

Sahai, A. K. (2016). Volatility modeling and forecasting efficiency of GARCH models on soy oil Futures in India and the US. *Journal of Energy Technologies and Policy*, *6*(3), 32–38.

Said, E., & David A. Dickey (1984). Testing for unit roots in autoregressive moving average models of unknown order. *Biometrika*, *71*, 599–607.

Sekhar, C. S. C. (2003). *Volatility of agricultural prices-an analysis of major international and domestic markets* (No. 103). Working paper, Indian Council for Research on International Economic Relations, New Delhi, India.

Tewari, H., Singh, H. P., & Tripathi, U. (2017). Growth and instability in wheat production: A region wise analysis of Uttar Pradesh, India. *International Journal of Current Microbiology and Applied Sciences*, *6*(9), 2537–2544.

Varshney, D., Roy, D., & Meenakshi, J. V. (2023). Impact of COVID-19 on agricultural markets: assessing the roles of commodity characteristics, disease caseload, and market reforms. In *Contextualizing the COVID pandemic in India: A development perspective* (pp. 249–271). Springer Nature Singapore.

Chapter 5

Role of Institutional Credit in the Development of Agricultural Sector: An Empirical Study on India

Sagnik Maity[a] *and Amit Majumder*[b]

[a]*University of Calcutta, West Bengal, India*
[b]*Bijoy Krishna Girls' College, West Bengal, India*

Abstract

The study aims to investigate the evolving dynamics of India's agricultural sector, particularly the shift from reliance on informal money lenders to a more prominent role of institutional credit in recent decades. This transition is seen as empowering farmers with increased access to funds for crucial investments in areas such as better seeds, fertilizers, irrigation systems and mechanization. Employing the Vector Autoregressive methodology on secondary data sources spanning from 1990 to 2022, the researchers seek to explore the impact of institutional credit on the agricultural sector. To address this, researchers used direct institutional credit and the share of gross value added of agriculture and allied sectors at current prices, finding that neither of them causes the other. Further analysis concludes that in the future, the contribution of credit will have some impact on development. Recognizing the significance of agriculture as the backbone of many economies, the study emphasizes the potential contributions of enhancing this sector through improved practices, advanced technologies and enhanced rural infrastructure. Ultimately, such developments are expected to drive increased productivity and overall output in the agricultural landscape, making a substantial contribution to economic growth.

Keywords: Institutional credit; agricultural sector; economic growth; VAR; cointegration; causality

Sustainable Agricultural Practices: Economic and Environmental Implications, 57–74

doi:10.1108/978-1-83608-336-820241005

Introduction

Agriculture plays a vital role in the Indian economy, serving as the backbone of livelihood for a significant portion of the population. It is not only a source of food and nutrition security but also a major driver of rural development and employment generation. India's agricultural sector employs over half of the country's workforce (Chand & Singh, 2022), providing livelihoods to millions of farmers and supporting various allied industries such as agribusiness, transportation and food processing. Additionally, agriculture contributes to foreign exchange earnings through exports of commodities like rice, wheat, spices and cotton (Ministry of Commerce & Industry, 2021). The sector also serves as a buffer during economic downturns, providing stability to the economy (Ministry of Agriculture & Farmers Welfare, 2023). Given its vast potential for growth and the increasing demand for food due to population growth and changing dietary patterns, the importance of agriculture in the Indian economy remains paramount, with ongoing efforts focused on enhancing productivity, sustainability and inclusivity within the sector.

In the fiscal year 2022–2023, India witnessed a substantial growth in agricultural credit disbursement, exceeding expectations and reflecting the government's commitment to address the financial needs of the sector (Ministry of Finance, 2023). The upward revision of the agricultural credit target for the following year underscores the government's determination to further enhance financial resources for farmers. Moreover, the focus on allied activities like animal husbandry, dairy and fisheries acknowledges the importance of diversification and value addition in agriculture to boost income and productivity in rural areas.

Efforts towards digitization and financial inclusion hold promise in extending credit access to underserved rural populations, potentially driving broader economic development and poverty alleviation initiatives. While informal credit has historically been accessible to farmers, it often comes with high-interest rates and exploitative practices, highlighting the need to promote institutional credit through formal banking channels (Pradhan, 2013). Government initiatives, such as those by the National Bank for Agriculture and Rural Development (NABARD) and regional rural banks (RRBs), aim to provide loans to farmers at subsidized interest rates, along with tailored financial services and schemes.

However, amidst these positive developments, pertinent challenges persist. Concerns regarding farmer indebtedness, though somewhat mitigated by government subsidies and interest rate support, necessitate vigilant oversight to ensure responsible lending practices and promote financial literacy among borrowers. Addressing regional disparities in credit access remains imperative to foster equitable growth across diverse agrarian landscapes. Moreover, the persistence of informal lending practices underscores the need for concerted efforts to integrate farmers into formal credit channels, thereby reducing reliance on high-interest informal loans.

This research aims to investigate the influence of institutional financial support on agricultural growth in India, with a specific focus on examining the relationship between institutional credit and agricultural development. Employing rigorous time-series methodology, the study endeavours to assess the efficacy of formal banking institutions in extending credit to farmers and agricultural enterprises. By scrutinizing the dynamics between institutional credit and agricultural development, the findings are poised to provide valuable insights into the pivotal role of credit in fostering sustainable growth within the agricultural sector. Moreover, the study's outcomes are expected to furnish policymakers with actionable recommendations aimed at enhancing the delivery mechanisms of credit within the agricultural domain, thereby facilitating more robust agricultural development initiatives.

Literature Review

Studies-Related Foreign Aspects

Institutional credit plays a crucial role in driving agricultural development by providing farmers with the necessary funds for investments and modernization, leading to increased productivity and output. A notable study by Balogun and Otu (1991) reviewed various credit policies and institutions aimed at stimulating agricultural development in Nigeria from 1979 to 1989. The work employed a simultaneous equation model to estimate the impact of credit policies on the supply and demand of agricultural credit, as well as the output elasticity of credit. Additionally, it tested the hypothesis of credit subsidies as production disincentives. The study found that credit policies had little influence on both lenders' and borrowers' behaviour and that credit had a limited effect on agricultural output. Furthermore, it revealed that credit subsidies reduced the profitability of agriculture and discouraged investment.

Similarly, research focused on Africa underscores the importance of credit policies in achieving sustainable economic growth, particularly in Sub-Saharan Africa (Abdulai & Hazell, 1996). Gylfason (2000) examined how natural resource abundance and agriculture affect economic growth in central and eastern Europe and central Asia. The working paper argues that dependency on natural resources and agriculture may stagnate economic growth. The author suggested that transition economies should decrease their dependence on natural resources and agriculture and enact legislative reforms to promote economic diversification, efficiency and sustainability.

Another study examines the influence of economic policies on sectoral composition and overall growth based on Argentina from 1913 to 1984. The work analyses macroeconomic, income and trade policies and examines the features of both the agricultural and non-agricultural sectors. The study offers insights into the impact of these measures on Argentina's growth path (Mundlak, 2000). Mundlak (2005) discusses economic growth from American agriculture, utilizing four periods of time series data. The study found that mechanization, labour productivity and technological changes played a key role in shaping agricultural growth. Additionally, Sial et al. (2011) conducted research in Pakistan, institutional credit on agriculture production. The study used Cobb Douglas production function using OLS method. The study indicates that agricultural financing, availability of water, yield of crops and labour force have a considerable impact on agricultural production.

Furthermore, Obilor (2013) reviewed an agricultural scheme in Nigeria and found that the scheme had a significant positive effect on agricultural productivity. These findings collectively emphasize the crucial role of institutional credit in fostering agricultural development and enhancing agricultural productivity across different regions. The study analyses the economic growth in developing countries within the field of agricultural economics. The research emphasizes the main areas affected by variations in growth rates, such as food demand, agricultural output and changes in structure. The change in demand towards animal goods and investments in agricultural advancement and research are important factors. Rural infrastructure and increased yields and productivity also have a role in overall economic development (Martin, 2019).

Additionally, a study conducted by Batmunkh et al. (2022) examines the effects of agricultural economic globalization (EG) on environmental sustainability across six Central Asian nations. The analysis includes aspects such as agricultural exports, value-added, currency rates, natural resource rents and foreign debt stocks. The work also examines the Environmental Kuznets Curve concept. While EG enhances economic growth and job opportunities, it also has negative impacts on the environment. The study underscores the importance of balancing economic development and environmental preservation in Central Asia.

Studies Related to Domestic Aspects

Numerous studies in India focus on the credit system and agriculture. Mohan (2006) conducted a study on the development of institutional agricultural credit from the 19th century, highlighting issues such as rural indebtedness, risk and exploitation by money lenders. The author found that while agricultural credit has grown significantly over the years, its share in total credit has declined. Additionally, studies conducted in India, such as that by Das et al. (2009), have examined the role of credit in agricultural production in India using dynamic panel data analysis. The study utilized 18 district-level panel data from 1999 to 2007, covering variables such as agricultural output, credit amount, credit accounts, fertilizer consumption, irrigation, rainfall, etc. The methodology used in the study was Arellano-Bond regression with instrumental variables to account for endogeneity, heterogeneity and serial correlation issues. The findings indicate that direct agricultural credit amount has a positive and significant impact on agricultural output, and its effect is immediate. Moreover, indirect agricultural credit accounts also have a positive significant impact on agricultural output, albeit with a one-year lag. Kumar et al. (2010) revealed that agricultural credit increased from 1972–1973 to 2008–2009. Employing Tobit model analysis, the authors identified factors influencing the quantum of institutional credit availed by farming households, including age, gender, household size, farm size, social group, education level and occupation type. The study suggested simplifying the loan procedure, capacity building of borrowers, strengthening special schemes for weaker sections and maintaining a balance between short-term and long-term credit to ensure sustainable agricultural growth. A study by Narayanan (2016) offers insights into the relationship between formal agricultural credit and agricultural GDP in India. Employing a mediation analysis framework and a control function approach, Narayanan finds that while credit has a positive and significant impact on input use, such as fertilizers, pesticides and tractors, its effect on agricultural GDP is weak and insignificant. In a comparable manner, Gulati and Juneja (2019) investigate the development of agricultural credit as well as the organizational efficiencies that have been discovered in the credit industry. The results of their investigation indicate that although credit has a favourable influence on input utilization, its influence on agricultural GDP is still very small and inconsequential.

Orhan et al. (2021) examined the correlation between carbon dioxide (CO_2) emissions and several economic indicators such as economic development, energy consumption, agriculture and trade openness in India. The study utilized data spanning from 1965 to 2019. This study employs a range of econometric methodologies to examine the relationship between CO_2 emissions and economic development, consumption of energy and agriculture in India. The findings reveal a positive connection, particularly in the medium- and long term, indicating that these factors have a detrimental impact on environmental sustainability in the country.

In contrast, Kumari and Gar (2023) employed an ARDL bounds test to analyse the effects of credit, interest rates and inflation on agricultural development from 1990 to 2019. They discover that while credit and inflation rates alone have a major impact in the short term, credit, interest rates and inflation all have positive long-term effects on agricultural development. The study also highlights challenges and opportunities for sustainable agriculture in India, including climate change, technological innovation and institutional reforms.

Collectively, these studies contribute valuable insights into the dynamics of agricultural credit and its implications for agricultural development in India.

Objective of the Study

The objective of the study is to investigate the impact of institutional credit on the development of the agricultural sector, exploring the causal relationship between these two variables. By analysing whether institutional credit causes development in the agricultural sector and vice versa, the study aims to elucidate the dynamic interactions between credit allocation and agricultural growth, providing valuable insights for policymakers, financial institutions and agricultural stakeholders.

Data Source

The research data has been sourced from the Database on Indian Economy (DBIE) and the KLEMS database published by the Reserve Bank of India (RBI). The timeframe of the data spans from 1990–1991 to 2022–2023, providing a comprehensive view of the financial landscape over several decades. The following variables are used in the study given below:

- ***Direct Institutional Credit to Agriculture*** refers to the direct lending provided to farmers and agricultural activities by various institutional sources such as co-operatives, Scheduled Commercial Banks and RRBs. This form of credit plays a crucial role in India's agricultural sector, as it provides financial support to millions of farmers, enabling them to invest in inputs like seeds, fertilizers, machinery and irrigation systems. The data for direct institutional credit to agriculture was collected from the DBIE in 2024 by the RBI. The data was transformed into a logarithmic series and denoted as 'ln_cre_agr' to facilitate analysis. This transformation allows for a more manageable representation of the data and aids in statistical analysis to understand the relationship between institutional credit and agricultural outcomes.
- ***The Share of Gross Value Added (GVA)*** of agriculture and allied sectors at current prices is a crucial indicator reflecting the contribution of agriculture and related sectors to India's economy. It signifies the proportion of GVA generated by agricultural activities, including forestry, fishing and livestock, concerning the total GVA of the economy. This data, collected from the KLEMS database in 2024 and published by the RBI, provides insights into the relative importance of agriculture within the broader economic landscape. The original series was transformed into a logarithmic series denoted as 'ln_gva_agr' to facilitate analysis and interpretation. This transformation enables researchers to examine trends, fluctuations and relationships with other variables more effectively, offering valuable insights into the dynamics of agricultural contribution to economic growth and development.

The consistent upward trend observed across all collected data points signifies a positive trajectory in the relevant metrics under examination.

Econometric Model

To determine the model, the first step is to check for stationarity. Stationarity is essential for time series analysis as it ensures that the statistical properties of the data do not change over time.

Hypothesis of the stationary test:

H0: The data series has a unit root and is not stationary.

H1: The data series has no unit root and is stationary.

Since both data series have *P*-values greater than 0.05 for tests including the intercept, trend and intercept and none model, the authors of the study fail to reject the null hypothesis, concluding that the data is not stationary. Given that neither of the variables is stationary at the level, the authors proceeded to check for cointegration test.

For the cointegration test the authors have done two tests: the Engle-Granger Cointegration test and Johansen Cointegration test.

Hypothesis of the cointegration test:

H0: There is no cointegration vector present in the equation.

H1: There is a cointegration vector present in the equation.

The authors of the study first conducted the Engle-Granger Cointegration test (Table 5.1), a statistical method used to determine whether two non-stationary time series are cointegrated, implying a long-term relationship despite being individually non-stationary. This test involves regressing one variable on another and examining the stationarity of the residuals. However, as the residuals of the test were not stationary at the level, the authors concluded that there is no cointegration vector present in the equation.

After conducting the Engle-Granger Cointegration test, the authors proceeded to perform the Johansen Cointegration test (Table 5.1), which is utilized to ascertain the number of cointegrating relationships among multiple non-stationary time series variables. This test estimates the rank of cointegration, representing the number of cointegrating vectors, through a likelihood ratio test. By comparing eigenvalue and trace value to critical values, researchers can determine the presence and strength of cointegration among the variables. However, in both test statistics, the *P*-value exceeded 0.05. As a result, the authors failed to reject the null hypothesis concluding that there is no cointegration present in the equation.

As no cointegration is found in the data, the vector autoregression (VAR) is found to be appropriate for the model. So, the econometric model is

$$\begin{aligned} ln\_gva\_agr = {} & \beta_0 + \beta_1 ln\_gva\_agr_{t-1} + \beta_2 ln\_gva\_agr_{t-2} \\ & + \beta_3 ln\_gva\_agr_{t-3} + \beta_4 ln\_cre\_agr_{t-4} \\ & + \beta_5 ln\_cre\_agr_{t-5} + \beta_6 ln\_cre\_agr_{t-6} + e_t \end{aligned} \tag{1}$$

Table 5.1. Cointegration Test.

Johansen Cointegration			**Engle-Granger Cointegration**		
Test based on	***P*-value**	**Accepted or rejected**	**Included in the test equation (residual)**	***P*-value**	**Stationarity**
As per maximum eigenvalue	*P*-value >0.05	H0: accepted	Intercept	*P*-value >0.05	Non-stationary
As per trace value	*P*-value > 0.05	H0: accepted	Trend and intercept	*P*-value >0.05	Non-stationary
			None	*P*-value >0.05	Non-stationary

Source: Authors' calculations.

$$
\begin{aligned}
ln\_cre\_agr = {} & \beta_0 + \beta_1 ln\_cre\_agr_{t-1} + \beta_2 ln\_cre\_agr_{t-2} \\
& + \beta_3 ln\_cre\_agrt_{-3} + \beta_4 ln\_gva\_agr_{t-4} + \beta_5 ln\_gva\_agr_{t-5} \\
& + \beta_6 ln\_gva\_agr_{t-6} + e_t
\end{aligned} \tag{2}
$$

Here, the t represents to the current year and $t-1$, $t-2$ and $t-3$ represent the lagged relationship.

Analysis of Findings

In this section, the authors followed a structured approach to analyse the relationship between variables in their model. Initially, they conducted a causality test to investigate the direction of causality between the variables. Subsequently, they proceeded to diagnose tests to assess the model's adequacy and reliability. Finally, they employed the impulse response function to visualize the impact of shocks in one variable on another, providing insights into the dynamic interactions between the variables. Additionally, the authors performed variance decomposition analysis to quantify the contribution of different shocks to the forecast error variance of each variable, enhancing our understanding of the relative importance of various factors influencing the system. This comprehensive methodology allowed for a thorough exploration of the relationships and dynamics within the model, providing valuable insights for interpretation and decision-making.

Causality Test

In this section, the authors have conducted several tests to analyse the relationships between variables. They utilized a VAR model, followed by a Block Exogeneity Wald test and a Pair-wise Granger Causality test.

VAR Model

In a VAR model, each variable is regressed on its own lagged values as well as the lagged values of other variables in the system, allowing for the simultaneous estimation of the effects of past values of all variables on their current values. After that the researchers performed the VAR Granger causality to measure the direction of the relationship.

Before performing the unrestricted VAR all the variables are converted into stationary and stored.

Table 5.2 presents the results of the vector auto-regression model, focusing on the association between two variables: *ln_gva_agr* and *ln_cre_agr*. The findings indicate a negative impact of each variable's own past values on its current values. Furthermore, the study reveals no significant impact of other variables on these two variables. In simpler terms, the authors have concluded that there is no discernible relationship between direct institutional credit and the development of agriculture based on the analysis.

Table 5.2. Causality Test.

Vector Auto-Regressive (VAR) Model			VAR Granger Causality/Block Exogeneity Wald Test		
Independent Variable	**Dependent Variable**		**Direction of Causality**	**Chi-Square Statistics**	**Probability**
	ln_gva_agr	ln_cre_agr	Share of gross value added (GVA) of agriculture → direct institutional credit to agriculture	1.174601	0.7591
ln_gva_agr (−1)	−0.804810** (0.20869)	0.002106 (0.44241)	Direct institutional credit to agriculture → share of gross value added (GVA) of agriculture	3.829933	0.2804
ln_gva_agr (−2)	−0.005888 (0.27334)	−0.038217 (0.57945)			
ln_gva_agr (−3)	0.250663 (0.21421)	−0.636714 (0.45411)			
ln_cre_agr (−1)	0.016604 (0.09255)	−0.513003** (0.19620)	Pairwise Granger Causality Test		
ln_cre_agr (−2)	0.010033 (0.09478)	−0.459827** (0.20093)	Null Hypothesis	*F*-Statistic	Probability
ln_cre_agr (−3)	0.089699 (0.08939)	−0.266343 (0.18949)	Share of gross value added (GVA) of agriculture Granger causes direct institutional credit to agriculture	0.39153	0.7603
Constant	−0.001419 (0.00903)	−0.004694 (0.01913)	Direct institutional credit to agriculture Granger causes share of gross value added (GVA) of agriculture	1.27664	0.3081
R^2	0.600224	0.427838			
Adj. R^2	0.486002	0.264363			

**Significance at 5% level.

Source: Authors' calculations.

VAR Granger Causality/Block Exogeneity Wald Test

Next the authors conduct the VAR Granger Causality/Block Exogeneity Wald Test (Table 5.2) and used to assess the direction and strength of causal relationships among the variables in the system.

Hypothesis for VAR Granger Causality/Block Exogeneity Wald Test:

H0: The lagged values of the ln_gva_agr do not Granger-cause the current values of the ln_cre_agr.

H1: The lagged values of the ln_gva_agr do Granger-cause the current values of the ln_cre_agr.

The *P*-value of the test statistics is greater than 0.05. This suggest that ln_gva_agr does not Granger-cause ln_cre_agr. This result implies that changes in share of GVA of agriculture do not reliably predict changes in direct institutional credit to agriculture.

Hypothesis for VAR Granger Causality/Block Exogeneity Wald Test:

H0: The lagged values of the ln_cre_agr. do not Granger-cause the current values of the ln_gva_agr.

H1: The lagged values of the ln_cre_agr. do Granger-cause the current values of the ln_gva_agr.

The *p*-value of the test statistics is greater than 0.05. This suggests that ln_cre_agr does not Granger-cause ln_gva_agr. This result implies that changes in direct institutional credit to agriculture do not reliably predict changes in the share of GVA of agriculture.

Pairwise Granger Causality Test

The authors have further conducted the Pairwise Granger Causality test (Table 5.2) to understand the validity of the studies as it provides a systematic framework for testing hypotheses about causal relationships, enhancing the rigour and credibility of research findings.

Hypothesis for Pairwise Granger Causality Test:

$H0_a$: Share of GVA of Agriculture dose not Granger causes Direct Institutional Credit to Agriculture.

$H1_a$: Share of GVA of Agriculture dose Granger causes Direct Institutional Credit to Agriculture.

$H0_b$: Direct Institutional Credit to Agriculture dose not Granger causes Share of GVA of Agriculture.

$H1_b$: Direct Institutional Credit to Agriculture dose Granger causes Share of GVA of Agriculture.

The authors further confirm both the variables are not Granger causes other variable, with both *P*-values exceeding 0.05. Consequently, the authors fail to reject the null hypothesis, affirming that the variables under study do not exert causal influence on each other.

Model Diagnostics

To validate the model, the researchers conducted model diagnostics, beginning with an examination of the *R*-squared values and the Durbin-Watson (DW) statistics. The *R*-squared values for both equations were found to be lower than the DW statistics, indicating that the regression is not spurious, as suggested by Bhaumik (2015).

Wald Coefficient Test

The first test performed was the Wald Coefficient test. It is used to assess the significance of individual coefficients and to determine which independent variables are important in explaining the variation in the dependent variable.

Hypothesis for Wald Coefficient Test

H0: All the coefficients are equal to zero or model does not exist.

H1: All the coefficients are not equal to zero or model does exist.

The value of the test statistics is less than 0.05. Therefore authors reject the null hypothesis and conclude that all the coefficients are not equal to zero or model dose exists.

AR Root Test

The second test conducted by the authors assesses the autoregressive (AR) roots of the model. This test is crucial as it ensures the stability and reliability of the model. A stable VAR model is characterized by having all AR roots lying inside the unit circle. If the model is unstable, it may produce residuals with non-random patterns, indicating a failure to capture the underlying dynamics of the data.

Authors have found that all the AR roots are depicted to lie within the circle, suggesting that the model is stable. This finding provides assurance regarding the reliability of the model and its ability to accurately represent the dynamics of the data.

Residual Test

This section contains the residual test of the model, which is crucial for assessing the adequacy of a regression model and ensuring that its assumptions are met. By examining the residuals, which are the differences between the observed values of the dependent variable and the values predicted by the model, researchers can identify potential issues such as heteroscedasticity, autocorrelation and non-normality of errors. Detecting these problems is essential because they can lead to biased parameter estimates, incorrect inference and unreliable predictions.

Autocorrelation LM Test

The first analysis conducted by the authors is the Autocorrelation LM test. This test aims to identify whether there is any autocorrelation or serial correlation present in the residuals of the model. Autocorrelation in the residuals suggests that the model may be incomplete or not adequately capturing the temporal dependencies in the data.

Hypothesis for VAR Residual Serial Correlation LM Tests

H0: No serial correlation present in the residual.

H1: Serial correlation present in the residual.

As the *P*-value associated with the Autocorrelation LM test is greater than 0.05, it indicates that there is no significant autocorrelation present in the residuals. This suggests that the model adequately captures the temporal dependencies in the data, enhancing its reliability.

Normality Test

Moving on to the second test, the researchers conducted a normality test using the Cholesky decomposition of covariance (Lütkepohl) orthogonalization method. If the residuals follow a normal distribution, it indicates that the model is appropriate for the data and the estimates derived from it are reliable. Conversely, significant departures from normality may indicate potential issues with the model's validity and may require further investigation or adjustments.

Hypothesis for Normality Test of the Residual

H0: All the residuals are normally distributed.

H1: All the residuals are not normally distributed.

As the *P*-value associated with the residuals, considered jointly, is greater than 0.05, the authors fail to reject the null hypothesis. This implies that all the residuals are normally distributed.

Homoscedasticity Test

The third test conducted by the authors assesses the homoscedasticity of the residuals. Homoscedasticity refers to the assumption that the variance of the residuals is constant across all levels of the independent variables. To evaluate this, the authors employed the White Heteroskedasticity test without considering cross-time dependence.

Hypothesis for Residual Heteroskedasticity Tests

H0: Residuals are homoscedastic.

H1: Residuals are not homoscedastic.

As the *P*-value associated with test statistics is greater than 0.05, it indicates that there is no evidence of heteroscedasticity, and the assumption of constant variance across observations is met.

Multicollinearity Test

The absence of high multicollinearity between the variables and their lagged values, as indicated by the authors' Multicollinearity test of the residuals, is a positive finding that strengthens the credibility of the regression model utilized in the study. Multicollinearity occurs when independent variables in a regression model are highly correlated with each other, leading to unreliable coefficient estimates and inflated standard errors.

H0: There is no multicollinearity among the residuals.

H1: There is multicollinearity among the residuals.

The authors' confirmation that the residuals are not excessively correlated suggests that the VAR model adequately captures the relationships between the variables under investigation. This supports the validity of the regression results and enhances confidence in the accuracy of the estimates derived from the model. Overall, the absence of high multicollinearity in the residuals bolsters the robustness and reliability of the study's findings.

Impulse Variance Function

The IRF is a valuable tool in VAR analysis that offers insights into the dynamic interactions and causal relationships among the variables included in the VAR model. By simulating the response of each variable to a one-unit shock in one of the variables while holding all other variables constant, the IRF reveals how shocks propagate through the system over time. This helps to assess the magnitude, direction and persistence of the effects of shocks on each variable, providing valuable information about the transmission mechanisms at play within the VAR system. By interpreting the IRF, researchers can gain deeper insights into the causal relationships and dynamic behaviour of the variables, aiding in the formulation of economic theories, policy implications and forecasting exercises. Overall, the IRF serves as a powerful analytical tool for understanding the complex interdependencies and dynamics inherent in multivariate time series data. Fig. 5.1 represents the impulse response of ln_cre_agr to ln_gva_agr innovation using Cholesky (d.f. adjusted) factors.

As the figure shows, there is a positive response of ln_cre_agr to ln_gva_agr shock which means that when ln_gva_agr increases, ln_cre_agr also increases. The response gap is getting smaller over the period. Therefore, the result suggests that the impact of ln_gva_agr on ln_cre_agr takes some time to materialize. The

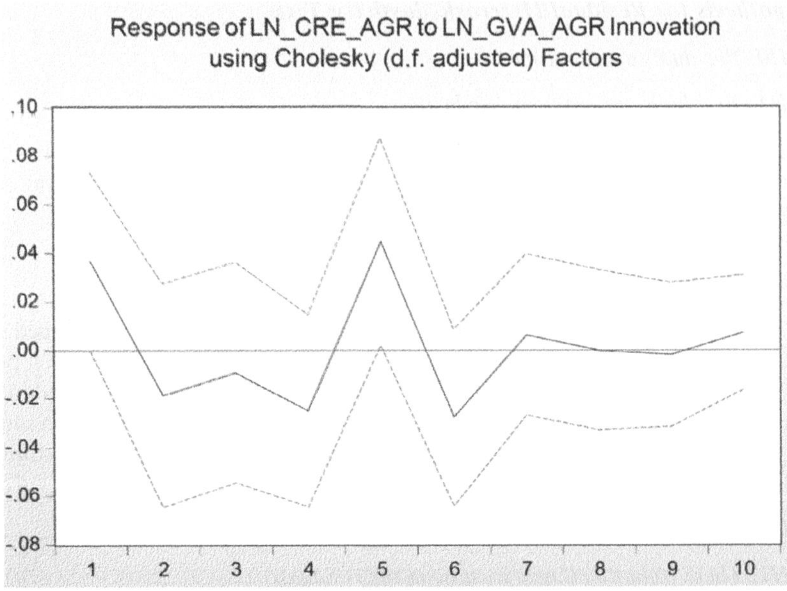

Fig. 5.1. Impulse Response Function. *Source*: Authors' calculations.

response peaks at around fifth period after the shock ln_cre_agr, and then the response gradually declines but remains positive for several periods.

Variance Decomposition Analysis

The variance decomposition analysis is a valuable technique in VAR modelling that provides insights into the relative importance of different shocks in explaining the variability of each variable included in the VAR system. By decomposing the forecast error variance of each variable into contributions from its own innovations (own variance) and innovations in the other variables (cross-variance) at various forecast horizons, the variance decomposition results offer a comprehensive understanding of the sources of variability within the system. This allows researchers to assess the extent to which each variable's behaviour is driven by its own dynamics versus the dynamics of other variables in the system. Interpreting the variance decomposition results helps researchers identify key drivers of variability, assess the degree of interdependence among variables and evaluate the transmission channels through which shocks propagate in the VAR model. Overall, the variance decomposition analysis provides valuable insights into the structure and dynamics of multivariate time series data, aiding in economic analysis, forecasting and policy formulation.

Overall, the standard errors in both variance decompositions came out relatively low, indicating a relatively high precision in the estimates. In the short run, variance decomposition of ln_gva_agr shows that 100% of the forecast error variance in ln_gva_agr is explained by itself, suggesting that the other variable in the model does not exert a strong influence. Specifically, ln_cre_agr exhibits strong exogenous influence in result, indicating that this variable is weak in the model. However, in the long run, the forecast error variance explained by the other variable increases slightly but remains very low. Conversely, in the short run variance decomposition of ln_cre_agr shows that ln_gva_agr explains 86% of the forecast error variance by itself, while 13% of the forecast error variance is explained by ln_cre_agr. However, in the long run, the influence of ln_cre_agr is expected to increase.

Conclusion

The study conducted a thorough investigation spanning 32 years to delve into the intricate relationship between institutional credit to agriculture and the development of the agricultural sector in India. Employing a VAR model, grounded on the stationarity of the data, researchers discerned that neither variable exerts a causal influence on the other. Subsequent rigorous testing through various residual examinations affirmed the robustness and validity of the model, with affirmative outcomes observed across all assessments.

Further analysis utilizing the IRF unveiled that shocks in either variable do not immediately reverberate onto the other, suggesting a nuanced temporal dynamic. However, over an extended period, the anticipated influence between the two variables gradually materializes. This discovery was supported by variance decomposition analysis, which illustrated an augmenting contribution of other variables' impact over the long term. Specifically, in the short run, each variable predominantly elucidates its own variance, while in the long run, the influence of external factors becomes increasingly pronounced.

The analysis sheds light on a concerning trend: despite the pivotal role of the agriculture sector in India's economic landscape, direct institutional credit has not wielded a significant influence on its development. This observation underscores the persistent inadequacy of institutional credit support, with informal channels retaining their sway. Indeed, as per a 2018 NABARD report, a staggering 52% of agricultural households in India found themselves in debt, primarily sourced from institutional avenues such as government schemes, cooperative societies and banks.

The existing discrepancies between credit providers and farmers continue to exist, which hinders the sector's potential for expansion. This is the case despite the abundance of programmes that are designed to support farmers. A progressive decrease in reliance on informal credit is projected to occur in the future, which is a glimmer of hope that can be seen in the near future. It is crucial that the government prioritize concerted measures aimed at restricting the informal credit system while simultaneously strengthening farmers' access to official loan channels in order to capitalize on this transition. This is important since it will allow the government to capitalize on this shift.

These findings offer valuable insights into the nuanced and evolving interplay between institutional credit and agricultural development in India. They underscore the significance of accounting for both short-term dynamics and long-term trajectories in formulating effective policies and guiding decision-making processes aimed at fostering sustainable agricultural growth and rural prosperity. By recognizing the intricate dynamics at play, policymakers can devise more nuanced and responsive strategies to bolster agricultural development and enhance the resilience of India's agrarian economy.

By enhancing accessibility to formal credit mechanisms, the government can usher in a new era of sustainable agricultural development while simultaneously alleviating the financial burdens weighing down farming households. Such a proactive approach not only fosters economic resilience within the agriculture sector but also contributes to broader efforts aimed at promoting equitable growth and prosperity across rural India.

Policy Recommendations

Based on the results of the study indicating the dynamic relationship between institutional credit and agricultural development in India, several policy recommendations can be proposed:

- **Enhanced Diversification of Credit Products:** To address the observed lagged impact of institutional credit on agricultural development, policymakers could encourage formal banking institutions to diversify their credit products tailored to the specific needs and constraints of farmers and agricultural enterprises. This could involve introducing innovative credit schemes with flexible repayment terms and interest rates, thereby catering to the diverse requirements of agricultural activities.
- **Investment in Agricultural Infrastructure:** Recognizing the long-term nature of the relationship between institutional credit and agricultural development, policymakers may prioritize investments in agricultural infrastructure. This includes improving irrigation systems, transportation networks, storage facilities and market linkages to enhance the productivity and market access of farmers. Such infrastructure development can complement the provision of credit by facilitating the efficient utilization of resources and reducing post-harvest losses.
- **Capacity Building and Financial Literacy Programs:** Given the importance of both short-term and long-term effects, efforts to enhance financial literacy among farmers and agricultural stakeholders could be intensified. Capacity building programs focusing on financial management, risk mitigation strategies and accessing formal credit channels can empower farmers to make informed decisions regarding credit utilization and investment in agricultural activities. Additionally, promoting awareness about government-sponsored credit schemes and subsidies can improve the uptake of institutional credit among the farming community.
- **Promotion of Agri-Entrepreneurship and Value Addition:** To leverage the increasing contribution of other variables in the long run, policymakers may incentivize agri-entrepreneurship and value addition initiatives. Encouraging the establishment of agro-processing units, cold storage facilities and farmer

producer organizations can create value chains that amplify the impact of institutional credit on agricultural development. Furthermore, supporting initiatives aimed at promoting sustainable agricultural practices and diversification into high-value crops can enhance the resilience and competitiveness of the agricultural sector.

- **Monitoring and Evaluation Mechanisms:** Establishing robust monitoring and evaluation mechanisms to track the utilization and effectiveness of institutional credit in driving agricultural development is essential. Regular assessments of credit disbursement patterns, repayment rates and the socio-economic impact on farmers can inform policy refinements and ensure the alignment of credit delivery mechanisms with the evolving needs of the agricultural sector.

By implementing these policy recommendations, policymakers can foster a conducive environment for leveraging institutional credit as a catalyst for sustainable agricultural development in India, addressing both short-term imperatives and long-term growth trajectories.

Limitations and Future Scope of the Study

In addition to the limitations mentioned, this study employs only two variables: direct institutional credit to agriculture and GVA contribution by agriculture. While these variables are insightful, incorporating additional variables could provide a more comprehensive analysis. Furthermore, the research makes use of the VAR model, which may be subject to the influence of structural breaks in the data. Structural breaks manifest as sudden and enduring changes in the underlying data-generating process, thereby influencing the relationships among variables.

References

Abdulai, A., & Hazell, P. B. R. (1996). The role of agriculture in sustainable economic development in Africa. *Journal of Sustainable Agriculture*, 7(2–3), 101–119. https://doi.org/10.1300/J064v07n02_10

Balogun, E. D., & Otu, M. F. (1991). *Credit policies and agricultural development in Nigeria*. Central Bank of India.

Batmunkh, A., Nugroho, A. D., Fekete-Farkas, M., & Lakner, Z. (2022). Global challenges and responses: Agriculture, economic globalization, and environmental sustainability in Central Asia. *Sustainability*, *14*(4), 2455.

Bhaumik, S. K. (2015, January 19). *Principles of econometrics*. Oxford University Press. https://india.oup.com/product/principles-of-econometrics-9780198098539

Chand, R., & Singh, J. (2022). *Workforce changes and employment some findings from PLFS data series*. NITI Aayog Government of India.

Das, A., Senapati, M., & John, J. (2009). Impact of agricultural credit on agriculture production: an empirical analysis in India. *Reserve Bank of India Occasional Papers*, *30*(2), 75–107.

Gulati, A., & Juneja, R. (2019). *Agricultural credit system in India: Evolution, effectiveness and innovations*. Center for Development Research of Bonn.

Gylfason, T. (2000). *Resources, agriculture, and economic growth in economies in transition* (CESifo working paper no. 313, pp. 1–50). CESifo Group Munich.

Kumar, A., Singh, K. M., & Sinha, S. (2010). Institutional credit to agriculture sector in India: Status, performance and determinants. *Agricultural Economics Research Review*, *23*(2), 253–264.

Kumari, A., & Garg, V. (2023). Impact of credit on sustainable agricultural development in India. *Journal of Sustainable Finance & Investment*, *13*(1), 560–571.

Martin, W. (2019). Economic growth, convergence, and agricultural economics. *Agricultural Economics*, *50*, 7–27.

Ministry of Agriculture & Farmers Welfare (2023). *Press information bureau*. https://www.pib.gov.in/PressReleasePage.aspx?PRID=1909213

Ministry of Commerce & Industry (2021). *Press Information Bureau*. https://pib.gov.in/PressReleasePage.aspx?PRID=1778071

Ministry of Finance (2023). *Ministry of finance year ender 2023*. Department of Financial Services.

Mohan, R. (2006, November). Agricultural credit in India: Status, issues and future agenda. *Economic and Political Weekly*, 1013–1023.

Mundlak, Y. (2000). *Agriculture and economic growth: Theory and measurement*. Harvard University Press.

Mundlak, Y. (2005). Economic growth: Lessons from two centuries of American agriculture. *Journal of Economic Literature*, *43*(4), 989–1024.

Narayanan, S. (2016). The productivity of agricultural credit in India. *Agricultural Economics*, *47*(4), 399–409.

Obilor, S. I. (2013). The impact of commercial banks' credit to agriculture on agricultural development in Nigeria: An econometric analysis. *International Journal of Business, Humanities and Technology*, *3*(1), 85–94.

Orhan, A., Adebayo, T. S., Genç, S. Y., & Kirikkaleli, D. (2021). Investigating the linkage between economic growth and environmental sustainability in India: do agriculture and trade openness matter? *Sustainability*, *13*(9), 4753.

Pradhan, N. (2013). *Persistence of informal credit in rural India: Evidence from 'All-India Debt and Investment Survey' and beyond*. MPRA Paper 80381, University Library of Munich, Germany.

Sial, M. H., Awan, M. S., & Waqas, M. (2011). *Role of institutional credit on agricultural production: A time series analysis of Pakistan*. Canadian Center of Science and Education.

Chapter 6

A Comprehensive Sustainable Entrepreneurship Model for Changing the Dynamics of the Market of Fertilizers in México During the Pandemic Crisis

José G. Vargas-Hernández[a] *and*
M. C. Omar C. Vargas-González[b]

[a]*Instituto Tecnológico Mario Molina Pasquel y Henríquez, Unidad Académica Zapopan, Zapopan, México*
[b]*Instituto Tecnológico de México, Campus Cd. Guzmán, México*

Abstract

This chapter aims to analyse the market of fertilizers in México during the pandemic crisis. The analysis of the fertilizer market in Mexico points out that the consumption of fertilizers has undergone a change in the structure in favour of consumers with the highest concentration and diversification. The analytical-descriptive and critical reflexive methods are applied both based on quantitative data gathered from secondary sources. This situation has contributed to a drop in the consumption of fertilizers because the farmer's real income has fallen during the pandemic crisis. To take benefit of the opportunities arising from the crisis, the functioning of an integral model of entrepreneurship in green innovation business (GIB) that is currently emerging and in the process of internationalization is analysed. With respect to the GIB and eco-efficient company through the comprehensive method of entrepreneurship, it is observed that just as the company has some advantages in the product, it is also with many threats that are the big companies that are already positioned and that also compete with

Sustainable Agricultural Practices: Economic and Environmental Implications, 75–91

Published under exclusive licence by Emerald Publishing Limited
doi:10.1108/978-1-83608-336-820241006

competitive prices. It is recommended that although the GIB has a wide network that is using the best way possible, it is necessary to increase the international network of contacts so that the business can grow quickly.

Keywords: Fertilizers; market; pandemic crisis; integral model; eco efficiency; internationalization

Introduction

The Mexican government's National Fertiliser Programme seeks to solve the issue of limited supply of national fertilizers at reasonable costs for small farmers. Includes chemical fertilizers and biofertilizers. This National Fertilizer Plan aims to reduce dependence on the import of these fertilizers. To achieve this, the Cosoleacaque Petrochemical Complex plants and the Pajaritos plant are reactivated to produce ammonia, an input to produce urea, which Mexico imports mainly from Ukraine. In the case of phosphates, in the Pacific, the Lázaro Cárdenas plant is operational and is the largest in Latin America. The National Biofertilizer Program shows incipient progress. The fertilizer production in Mexico estimated for 2019 was 1.85 million tons, reflecting an annual reduction of 2%; while demand continues to rise, with a record estimate of around 5.5 million mt. However, by July 2020, the production volume of nitrogen fertilizers in Mexico almost reached 48,800 metric tons, which represents a decrease of 32.7% compared to that reported during the same month in 2019. The production volume of phosphate fertilizers in Mexico exceeded 75,600 metric tons, which represents a decrease of 29.1% compared to that reported during the same month in 2019 (Salas, 2020).

Currently, there is a new business model where the objective is to be eco-efficient. Eco-efficiency is defined as the production of products and services at competitive prices that meet human needs and provide quality of life, while the ecological consequences and the use of numerous resources during the life cycle are progressively reduced and level equivalent, at least, to the estimated capacity of the planet (World Business Council for Sustainable Development, 1991).

In contrary, it is mentioned that eco-efficiency has the purpose of establishing the production of manufactured products of high durability, reducing the intensity in the application of energy for the production of goods and services, maximizing the use of raw materials, managing and disposing of hazardous materials and waste in an efficient and environmentally acceptable manner, have management systems and environmental quality, as well as procedures in occupational safety and health, among other provisions, that will bring them financial benefits and competitiveness (Cantú, 2008, p. 78). In both definitions, the authors agree that eco-efficient companies should have as their main objective, to develop quality products at competitive prices, as well as to reduce the environmental impact of producing or offering their products and services. Castro (1998) mentions that eco-efficiency aims to address three related aspects that correspond to (1) the total quality, which involves productivity and quality in the company; (2) the preservation of the environment, which is related to sustainable development; (3) occupational health and safety (Castro, 1998).

Fertilizer Demand in Mexico

A recent analysis of the fertilizer market in Mexico has pointed out that the consumption of fertilizers has undergone a change in the structure in favour of consumers with the highest concentration and diversification (UACH). This situation has contributed to a drop in the consumption of fertilizers because the farmer's real income has fallen. On the change in the consumption pattern (sources), I have no hard information about it. From experience in the field, it can be quantified that there has been growth in the use of physical fertilizer mixtures where the distributor makes certain formulations, mixing fertilizers, and thus the producer only buys one product. Regarding high concentration, it is well understood, if probable, that there has been a significant growth in protected agriculture such as strawberries and high-value vegetables where drip irrigation and fertilizers are used, and highly soluble sources with minimal impurities are regularly used, which are not normally traditional sources.

However, with more current data reported by Instituted Trusts in relation to Agriculture (FIRA 2020), in the agricultural year 2019, 71.8% of the sown area was fertilized, representing the seventh consecutive year with increases in the percentage of fertilized area. According to this report, the consumption of fertilizers in Mexico grew 5.8%.

In an interview with an FIRA fertilizer specialist analyst, Gallegos Cedillo (2021) asking about the trend in fertilizer consumption and about the demand and supply of fertilizers in Mexico, he warned that the lack of information with hard data does not always support trends in which the market moves. Regarding the consumption of fertilizers, the production information indicates that production has fallen: 15.4% from 2013 (2.06 million tons) to 2020 (1.75 million tons) for the period January–September. This, although in 2014 and 2019, increased compared to the previous year and the same period.

Now, if we consider the fertilized area, it has increased, as a percentage of the sown area. This proportion has augmented from 2013 to 2019, the last reported year, going from 65.3% in 2012 to 71.8% in 2019. However, if we look at the apparent consumption, it increased from 5 million tons in 2013 to 6.7 in 2018; in 2019 (January–December) it does drop to 5.8, and in September 2020 it is 5.2. There are no elements to attribute the drop in consumption from 2018 to 2019 to the decrease in producers' income, but if someone affirms this, it is difficult to prove otherwise.

Fertilizer Production in Mexico

The increase in Mexican fertilizer production during the last years is attributed, in part, to the reforms made to the Pemex Law since 2008. These had the objective of promoting the production and productivity of the Mexican fertilizer industry through the supply of raw materials, such as ammonia, at competitive prices for national manufacturers. In 2018, national production contributed 30.7% of apparent national consumption, which is estimated at 6.2 million dollars and represents an annual increase of 7.9%. The foregoing shows the country's high dependence on fertilizer imports.

With data from 2019, the national production of fertilizers is mainly composed of phosphate fertilizers (diammonium and others) with 55.7%. Nitrogen fertilizers (ammonium sulphate and nitrate and others) with 32.2% and acidic fertilizers (phosphoric, sulphuric, and nitric) with 10.9%. The estimated value of the fertilizer industry in Mexico in 2019 is 13,616.4 million pesos, 2% lower than that registered in 2018. Fertilizer production processes in Mexico are not integrated, a situation that has an impact on domestic producers, especially urea and nitrate, being at a disadvantage compared to producers of international competition (UACH). According to data from the latest Quarterly Report on Fertilizers from the Directorate of Research and Economic Evaluation of the Instituted Trusts in relation to Agriculture (FIRA, 2020), the production of fertilizers in Mexico decreased at an annual rate of 8.4% in the first nine months of the year compared to the same period in 2019.

Regarding the fact that the national producers of urea and ammonium nitrate are at a disadvantage with international competition, as a result of the lack of integration of their production processes, if the statement is well understood, especially the reason that is manifested by the lack of integration of their production processes, in general, it would be expected that greater integration greater profitability, but this is not measured to be the main cause for the disadvantage of domestic producers. It can affect, but it is not the main reason.

More weight would be given to the characteristics of the Mexican farmer, small areas and little specialized; he grows various products, as well as the dependence to a large extent on the import of fertilizers that expose him to have fertilizers without due quality control, lack of weight in the containers (49 kg bags instead of 50, e.g.), exchange rate, etc. (Gallegos Cedillo, 2021).

Fertilizer Exports and Imports in Mexico

In this regard, with information from the Ministry of Economy, imports of fertilizers in Mexico, during 2018, registered a volume of 5.06 million dollars and represented an annual increase of 12.6 percent. In the same year, imports came from Russia (30.9%), the United States (14.3%), Norway (12.1%) and China (11.1%) mainly. In the last five years, these countries participated with 69.5% of the national fertilizer imports. In the case of exports, these stood at 0.79 million pesos and registered an annual growth of 10.5%. Between 2008 and 2018, consumption and imports grew on average at an annual rate of 6.3 percent. Imports in 2019 are estimated at around 2.9 million tons, of which an advance of 92% was recorded as of November. Of the total imported, 67.2% corresponded to nitrogenous, 15.5% to phosphate, 12.5% to potash and others with 4.8%. The commercial value of these imports was 807.8 million dollars, reflecting a reduction of 14.3% in relation to the record of 943.1 million dollars registered in 2018.

The main supplying countries were Russia 30.6%, China 16.9%, United States 11.3%, Algeria 7.9%, Malaysia 3.8%, Chile 2.9%, Canada 2.1%, Egypt 1.4%, and others 23.1%. Between January and September 2020, Mexican fertilizer imports grew 9.9%, while exports decreased 16.6% at an annual rate, totalling 3.9 and 0.46 million dollars, respectively.

The Fertilizer Market

The implicit price of fertilizer imports in Mexico in 2018 was $322.3 per ton and represented an annual increase of 6.7%. The implicit price of imports from Russia was 283.4 dollars per ton, 387.2 for those from China and 416.1 for those from the United States and showed an annual growth of 10.8, 24.2, and 9.7%, respectively. Regarding the average price level of fertilizers in Mexico, since 2014, it has been on the rise. In 2018, a ton of fertilizer, in distribution centres, was quoted on average at 10,254 pesos, which represents an increase of 3.5% compared to 2017. The fertilizers with the highest annual increases in prices were diammonium phosphate (DAP) (10.4%), triple 17 (10%) and potassium nitrate (9.4%). In January and February 2019, the average price was reduced by 2.0 and 1.3% compared to December 2018, so it would be expected that the average price in 2019 will be at levels like those registered during 2018. In the domestic market, urea prices increased 7.3% from 2018 to 2019; while those of DAP rose 4.8%.

The average price of fertilizers in the country maintains an upward trend. In November 2020, it registered growth of 2.7% at an annual rate and 3.3% from December 2019 to date. As of November, the price increases of triple 17 (12.2% annually), ammonia (9.9%), ammonium sulphate (8.3%) and ammonium nitrate (7.9%) stand out. The prices of simple superphosphate and potassium chloride were the ones that showed the greatest annual decrease, at rates of 7.1 and 4.6%, respectively.

Challenges and Opportunities: A Comprehensive Entrepreneurship Model of Green Business Innovation

Fertilizers are an input of fundamental importance in primary agricultural activity. The rehabilitation of fertilizer producing plants in the country is not economically viable in the short term because the reactivation of activities in the plants takes time, it is necessary to invest more resources for the qualification and the import of ammonia gas continues, an input necessary in the production of fertilizers. The importation of fertilizers will continue since the national production of 2020 only covers 33.7% of the total demand, and the remaining 66.3% is brought from the foreign market. Entrepreneurship, inclusiveness, civil and culture are variables that have complex but straightforward linkages and processes among them and much is still unknown. In recent years, there has been a significant amount of theoretical and empirical research on entrepreneurship and its influence on regional and business economic performance (Evans & Leighton, 1989; Holmes & Schmitz, 1990). However, when linking entrepreneurship to the variables of inclusiveness and civil culture, the literature is inexistent. There is a relatively void, despite some recent efforts to study the relationship between entrepreneurship in some other related fields beyond economic growth, which may be attributable to the lack of theoretical frameworks.

However, some studies conducted in economic development have shown reversed causality influencing entrepreneurial activities (Ortega-Anderezand & Lai, 2017). Entrepreneurial activities tend to be overestimated in regions where there are start-ups playing a relevant role while entrepreneurship activities tend to be underestimated in regions where there is a new entrepreneurial formation and the start-ups are relatively few (Baptista et al., 2010). Researchers have focused on the significance of entrepreneurship in economic growth, demonstrating how it affects employment and social development. Entrepreneurial intentions promote entrepreneurial behaviours (Krueger et al., 2000; Souitaris et al., 2007; Veciana et al., 2005). Entrepreneurship activity can be associated with human necessity providing an opportunity to earn money for living, until better alternative opportunities are found on the labour market (Carree & Thurik, 2010).

The concept of entrepreneurship is multidimensional and related to individual willingness, abilities and activities on their own, in teams and in organizations to make decisions facing obstacles and uncertainty on the use of institutions and resources to create and develop new opportunities (Wennekers & Thurik, 1999). Entrepreneurial opportunities are the result of unemployment the influences start-up activity, the consequence of a thriving economy and experience in entrepreneurial activities (Lin et al., 1998; Pfeiffer & Reize, 2000). Entrepreneurship is the recognition and exploitation of opportunities leading to the creation and development of a firm (Aragon-Sanchez et al., 2017). Moreover, entrepreneurial intentions predict entrepreneurial behaviours according to the planned behaviour theory (Ajzen, 1991). Entrepreneurial activity creates opportunities to influence economic performance (van Stel et al., 2005) such as entering the markets with new production processes and products (Acs & Audretsch, 2003). Among the entrepreneurial models there is evidence that the entrepreneurial intentional models (Krueger et al., 2000) support the theory of planned behaviour.

The theory of planned behaviour has been used in the analysis of entrepreneurial intentions (Shook et al., 2003). The determinants have significant positive association with behaviour intention (Armitage & Conner 2001). Do Paço et al. (2011) found a positive and significant influence between entrepreneurial intentions, attitudes and perceived behaviour control. However, subjective norms have an indirect impact on entrepreneurial intentions. The determinants of entrepreneurial intentions are linked with access to financial, natural, human and cultural capital and resources through the influence of attitudes, individual subjective norms, perceived social control and self-efficacy. The attitude is the individual favourable evaluation to start a new opportunity. Individual subjective norms are the perceived social pressure to start a new opportunity and perceived social control and self-efficacy is the perceived ease to start a new opportunity. Based on the Shapero and Sokol's (1982) model, attitudes and subjective norms are linked to desirability and perceived behavioural control or self-efficacy with feasibility (Krueger et al., 2000).

Methodology of Research

Background of the Problem and Assumption

Green innovation in urban areas is a neglected issue in terms of urban planning and policing. Still, more neglected is a concern for changes in urban green areas towards the implementation of green innovation initiatives to revitalize the cities, increase the economic growth, improve the social justice and inclusiveness, as well as the improvement of environmentally sustainable development, strengthen the biodiversity and socio-ecosystems. To achieve these goals, it is necessary to implement some actions following the design and implementation of a comprehensive entrepreneurship model. This new eco-efficient business entrepreneurship model will be analysed in the context of the integral model, analysing in this way the tripod of the strategy, which integrates considerations based on industry, resources and institutions. This chapter begins with the assumption that the companies of ecological base present major difficulty at the time of wanting to internationalize in comparison to the companies of the industrial base.

Instrument and Procedures

This analysis is based on the specific case study of an eco-efficient company using the methodological tools proposed for each one of the theoretical approaches. For the analysis industry-based considerations are employed in the model of the five forces, also called the diamond model proposed by Porter (1980). The resource and capabilities analysis of the eco-efficient company is employed, and the model proposed by Barney (2001) complemented with the SWOT analysis. Finally, the eco-efficient company is also analysed in terms of the institutional considerations considered as the 'rules of the game', formal and informal, that frame the entrepreneurial activities and the creation and development of companies according to the territorial environment.

Data Analysis: Use of Fertilizers in Mexico

The National Development Plan 2019–2024 (https://www.gob.mx/senasica/documentos/agriculture-prioritizes-health-and-safety-actions-in-2024) establishes among the priority actions for food self-sufficiency and the rescue of the field the fertilizer programme for the benefit of agricultural producers.

Bio Organic Fertilizers in México

The use of organic fertilizers in Mexico is not very common; this is because industrial-based companies need chemical products that make plants grow at a faster rate due to market demand. But these fertilizers damage the soil causing it to become unusable after a time for the harvest; as a result, the food absorbs these fertilizers that are harmful to the human being in the long run. On the other hand, organic fertilizers bring many benefits to the soil along with plants and food. According to a study carried out by SAGARPA, organic fertilizers favourably influence the physical characteristics of the soil (physical fertility); These characteristics are structure porosity, air action, water retention capacity, infiltration, hydraulic conductivity and stability of aggregates. Table 6.1 shows a comparison made by the National Agricultural Survey (ENA) in which there is a rise in the use of chemical fertilizers in Mexico and a reduction in organic fertilizers.

Theoretical and Conceptual Framework

The theoretical framework used in this study is based on analysing the coefficient firm using the major three theories of strategic design and implementation. The industry-based considerations, the resource and capabilities considerations, and the institutions-based considerations are shown in Fig. 6.1.

The Green Innovation Business (GIB): BIO-FOM

BIO-FOM is a green innovation business (GIB) also characterized as an eco-efficient company to promote sustainable, profitable and inclusive development through highly competitive Mexican seeds with fair prices. Using organic mineral bio-fertilizers healthily increased the profitability of the producers. It is in the metropolitan area of Guadalajara.

Table 6.1. Difference of the Use of Chemical and Organic Fertilizers in Mexico National Agricultural Survey.

Agricultural Technology (First Part)		
Employed Technology	%	%
	ENA 2012	**ENA 2014**
Type of seed	60.9%	82.2%
Creole	29.7%	29.2%
Improved	NA	NA
Certified	NA	NA
Transgenic	NA	NA
Seedling	NA	21.0%
Chemical fertilizers	65.5%	68.8%
Natural fertilizers	40.4%	27.5%
Herbicides	61.7%	62.7%
Insecticides	45.3%	48.2%

Source: Author's deliberation using INEGI; National Agricultural Survey (ENA, 2014).
Note: The sum does not give 100 because each production unit can practice more than one technology; NA: not available.

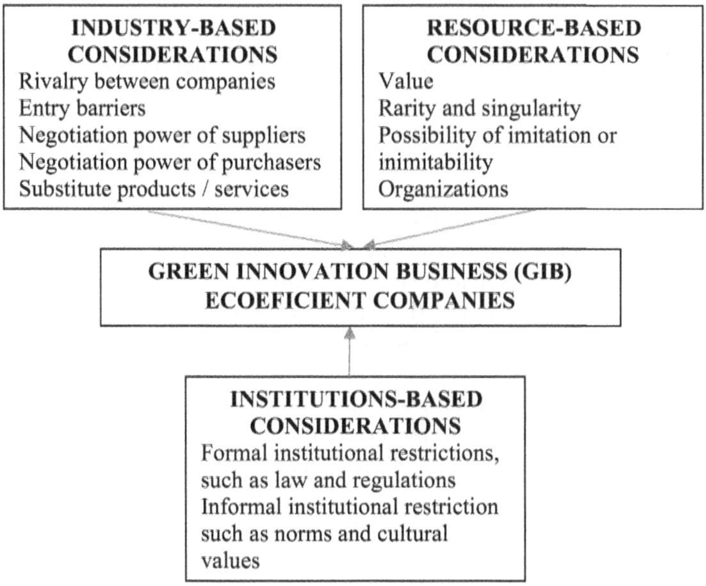

Fig. 6.1. A Comprehensive Entrepreneurship Model for the Internationalization of Green Innovation Business (GIB) and Eco Efficient Companies. *Source*: Own elaboration based on Peng (2012).

BIO-FOM is the most complete organic mineral bio-fertilizer available on the market, which is made from the interaction of elements: biological, organic and mineral. It is a mineral organic bio-fertilizer for plant nutrition. The interaction of the BIO-FOM elements forms a Functional Plant Nutrition System; its results reflect: improving seed germination and initiation, healthy and adequate growth, larger roots, greater flowering and tie increase in the quality of the fruits and provide resistance to pests and diseases. BIO-FOM increases soil fertility and contributes to the decontamination and regeneration of the soil.

BIO-FOM bases its functionality on the interaction of the biological, organic and mineral elements that compose it: A Poly-Functional Consortium of Microorganisms, among them: antagonists to pests and diseases, nitrogen fixers and mycorrhizal fungi that help release the nutrients provided by the BIO-FOM blend itself. The 17 basic mineral elements of plant nutrition (nitrogen, potassium, phosphorus, calcium, sulphur, magnesium, iron, boron, copper, manganese, silicon, zinc, iodine, sodium, cobalt, molybdenum, selenium) organic matter, amino acids, humic acids and fulvi.

BIO-FOM contains live microorganisms: eight bacteria and mycorrhizal fungi that help unblock the chemical components that exist in the soil allowing their absorption, in addition to an important contribution of organic matter, amino acids, humic acids and fulvic acids; this, together with its content of 16 mineral elements (nitrogen, phosphorus, potassium, magnesium, calcium, iron, boron, copper, manganese, silica, zinc, iodine, sulphur, sodium, cobalt and chlorine) give rise to a molecular dialogue that results in a functional plant nutrition system, fertile soil, larger root size and helps improve pH, which is essential for proper and healthy plant growth.

Among other benefits, it potentiates the chemical fertilizers that complement the application of BIO FOM as:

a) It recovers the soil fertility, unlocking mineral nutrients present but not available, favouring their absorption due to their bio-organic profile.
b) Increases the specific weight of grains, resulting in higher production.
c) Increases the Br ix degrees of the fruit, having better sales possibilities.
d) Stimulates the immune system, so healthy, vigorous and productive plants are achieved. Strengthens the root system of the plant, thereby improving the use of water and mineral nutrients available in the soil.
e) Respect the ecosystem, avoiding the contamination of the water table.
f) 100% assimilable, achieving a greater expression of the genetic potential of plants.
g) Retains and conserves moisture in soils.

Empirical Results and Conclusions

Industry-Based Considerations

When analysing industry-based considerations, the framework of Porter's five forces is used, which considers factors such as rivalry between firms, barriers to entry, the power to negotiate with suppliers and buyers as well as substitute products.

The Rivalry Between Companies

When talking about the rivalry of companies, it refers to the struggle that companies face every day to obtain a position before consumers. These battles can be for prices, quality in products, advertising battles, etc. In contrast, the author Huyghebaert mentions that the rivalry between firms generates a direct influence on the likelihood that a business will be successfully undertaken, as well as having an effect on the existence of barriers to entry since fewer firms exist in one industry more complicated will be the entry of new firms (Huyghebaert, 2004).

The company to analyse is devoted to organic fertilizers made from organic minerals. Because it is a new product, it does not face much rivalry; nonetheless, there is competition from alternative goods, which are all chemical fertilizers, due to the wide range of substitute products accessible in the marketplace. It is difficult to compete for price in the same way as being a medium-sized company is not even positioned in the consumer's mind when thinking of organic fertilizer.

Nowadays fertilizers and other chemical products for the treatment of the soil represent 15.7% of the income of the nurseries and gardens industry in 2019.

The sector has grown rapidly in recent years, owing to rising disposable income per capita and a stronger consumer segment. Scotts Miracle-Gro Company, Walmart and Lowe's have dominated a sizable portion of the US market. Because economies of scale, as well as large corporations, can maintain a low cost while offering affordable prices, price rivalry among major corporations has resulted in weak industry development and low profit margins. Some less profitable corporations were compelled to exit the industry. Successful businesses are able to compensate for high pricing by providing a better understanding of the product and excellent customer service.

Over the next five years, the nursery and garden store industry are expected to continue to grow, as economic conditions are expected to remain stable. As the disposable income per capita increases, the consumer will be able to invest more time and money in the appearance and operation of their gardens. It is also expected that the industry will benefit from a growing demographic interest in small plants, gardening and organic horticulture.

Below is a list of companies with which Mary's Poop competes in the United States.

Scotts Miracle-Gro Company (OH, USA)

Founded in 1868, Scotts Miracle-Gro Company is a multinational corporation and industry leader in the lawn and garden market. Its products include the infamous Miracle-Gro and a complete line of plant foods. The most popular Miracle-Gro fertilizers are inorganic, which are much less expensive than organic fertilizers.

PacaProGro (CA, USA)

The PacaProGro farm is in Somis, California, with an estimated 180 alpacas producing alpaca fibre and alpaca manure. It not only sells alpaca fertilizers but also merchandises products made of alpaca fibre, such as hats, gloves, scarves, and stuffed animals.

Alpaca Grow (VT, USA)

Alpaca Grow is a small family business operated in Vermont that began in 2014. Like Mary's Poop, Alpaca Grow sells only alpaca fertilizers.

Isolated Wood Farm (British Columbia, Canada)

Secluded Wood Farm is headquartered in British Columbia, Canada. They sell alpaca fibre and manure products.

Alpaca Island Company of Martha's Vineyard (MA, USA)

Island Alpaca Company has more than 15 acres and around 50 alpacas. They sell alpaca fibre and manure products, and they charge higher prices.

Camelot Haven Alpacas (British Columbia, Canada)

Founded in 2002, Camelot Haven Alpacas produces and sells alpaca clothing, accessories, gifts and alpaca fertilizers.

Entry Barriers

As already mentioned before, BIO-FOM faces competition from large chemical and organic fertilizer companies. Due to this, there are problems when it comes to wanting to enter the foreign markets, as the big brands are doing well positioned in the market, and it would be complicated to unseat them or compete for prices so that the company, in this case, could compete for product quality as well as innovation and for the benefits that its product provides.

It is also significant to consider the transaction cost and the transportation cost if working with foreign currencies. If the costs become too high, BIO-FOM products would be less competitive in the foreign market. An alternative to export is the franchise, which allows local business units to produce BIO-FOM products for a fee.

One of the competing companies of Mary's poop is already positioned in 10 states of North America and has large sales volumes, being this way, the products of Mary's poop and other small alpaca farms cannot compete by prices and volume, as already mentioned previously, which makes it be handled as an oligopoly market among large companies. These in turn set entry barriers when playing with prices since, if they see that a new product wants to enter the market, they lower their prices to continue to have an advantage over others or use promotions which make them more attractive towards the purchase of their products to the final consumer.

It is also central to consider the transaction cost and the transportation cost if working with foreign currencies. If the costs become too high, Mary's Poop products would be less competitive in the foreign market. An alternative to export is the franchise, which allows local alpaca farms to produce Mary's Poop products for a fee.

Substitute Products or Services

In the global marketplace and more specifically, North America, large, medium-sized and small competitors produce similar products. Geographically, most of the production plants are in rural areas and seem to serve only local markets. The competitive advantage of BIO-FOM is that it is the company with the highest social awareness among all its competitors. And although its product is easy to imitate, it gives it a plus by mixing it with some other ingredients that make the compost a better quality. It is also easy to use as they pack the product inside disposable pod-made biodegradable materials that only must be deposited in a container with water waiting for it to dissolve and start to water in a normal way. But there is a high range of substitute products that, although they are not organic, the final use is the same (serve as fertilizer to the plants), sharing with substitute products is complicated since the raw materials they are made with are very economical and this it makes the prices of the products accessible to all types of publics. On the contrary, the organic products that Mary's poop handles are of a slightly higher price, and this is due to the handmade process with which they are manufactured.

Bargaining Power with Buyers

According to Porter, at this point, it is defined as the ability of customers to impose prices and conditions of sale (Porter, 1980). This force can be established by customers directly, whether negotiating a discount or financing model, demanding delivery forms, or indirectly which is summarized with competitive purchases. The bargaining power of buyers may depend on some variables such as a high supply of products and/or services and low demand for them. Another may be that the products offered have no differentiation among themselves, etc. At this point, the customer has the option to choose any product or service that is presented to them and that they consider to be the best and meet their expectations. As well as defining what is the maximum price customers are willing to pay for a product or service, as well as some other requirements that could be delivery times, product quality, etc. All this has an effect on the company's profits.

Mary Forte, the owner of Mary's Poop, manages and operates the company and interacts with customers and other interested parties through social networks. Nowadays, online/digital marketing is the most important marketing tool, especially in social networks. For example, use 'Pinterest' to communicate with the company's audience. This social networking platform is popular with many users who share images that can be easily shared. Therefore, she can approach her audience and attract their attention and thus know what customers need and adapt to the needs of the market.

In one of his writings, Peng mentions that when the bargaining power of suppliers becomes too great, business solutions must be found that can reduce it (Peng, 2012, p. 127); this is since many options must be available to purchase inputs at reasonable prices. The organic fertilizer based on organic minerals requires many inputs because a lot of raw materials are used for its elaboration, so it is necessary to have many suppliers. But if it is necessary that with few suppliers that have contact make and establish the negotiation agreements and delivery times among many other things, they can reduce and, to a certain extent, eliminate transaction costs.

Considerations Based on Resources and Capabilities

The resources are any input in a productive system in which an output is generated. These can be classified as financial, physical, human, technological, organizational, knowledge, management team experience and customer service, among others. The consideration based on resources groups some indispensable factors for entrepreneurship, some of them are creating value, being unique through the rarity factor among other aspects. For example, in an analysis Peng mentions that it is important to consider the resources of the company; they must be oriented to create value, they must have rarity, they must be difficult to imitate and finally business resources must be immersed in the organization (Peng, 2012). On the other hand, Barney dogmatizes that the heterogeneity of organizations is due to the possession of resources: (i) valuable, which must respond to environmental threats and take advantage of their opportunities; (ii) rare or scarce, those that cannot allow obtaining competitive advantages with competitors; (iii) difficult resources to imitate, without substitutes and organizational, which means that the company has aspects of order (Barney, 2001, p. 41).

The company BIO-FOM offers the market an innovative product because it is taking the greatest benefit to a product that people see as a waste and transformed it into a product with added value; the strengths which this product provides are that they are organic, have a higher performance compared to other fertilizers, are favourable to the environment and are not expensive to produce. Something that gives a higher value is that the packaging is biodegradable by which makes the whole product itself is ecological and there is no waste of plastic or any other material that is difficult to decompose, the product is safe for children and pets so anyone can use it. As a result, for resource considerations, a SWOT analysis was performed to analyse both internal and external elements that bring value to the organization, as well as positive and negative aspects that may lead the company to not develop as quickly as projected (Table 6.2).

In this table, it is shown that the firm has many strengths that make its product different but also has many threats. This is due to the same rarity of the product that makes it somewhat complicated for sale; likewise, the product is easy to imitate, so at any time it could have a lot of competition, which can quickly get to the market firm. In his essay, Porter states that the competitive advantage comes from the value that a company manages to create for its customers by offering special benefits that compensate a higher price of the product (Porter, 1980).

Considerations Based on Institutions

When talking about considerations based on institutions, we are talking about the rules of the game of these, in which the behaviour of the company is determined and how they are developed around the world. Peng points out five strategies for an entrepreneurial company to be successful, which can be applied together, and they are growth, innovation, networks, financing/government and harvest/output (Peng, 2012).

The BIO-FOM company incorporates some of these strategies such as innovation in their products, as well as networks because they try to have a wide network to get known as well as to obtain advice and keep growing as the company participates in a programme called X-culture where enterprises are assigned a group of people from different parts of the world, advising companies so they can expand or internationalize at the same time the company uses all the networks that are possible. Peng mentions that there is evidence that networks, personal and organizational, represent significant resources and opportunities and that the successful creation of networks can lead to the efficient performance of firms (Peng, 2012, p. 13). Instead, McDougall notes that network analysis builds

Table 6.2. SWOT Analysis.

Strengths	Opportunities
Organic products/products without chemicals	Growing trend in organic products
Excellent performance compared to other inorganic fertilizers environment friendly	Market development
It is not expensive to produce	Product development
The packaging is biodegradable	Product differentiation
No toxic, safe for children and pets	
It is not easy to imitate	
Weaknesses	**Threats**
Lack of advertising + visibility	Regulations that exist for exports
Small production capacity-challenge for the company at scale	Animal diseases
The products have limited functions	Environmental factors may affect production
The products are not standardized	Breach of demand levels
Hard to forecast production	Low market level
Limited capital and strategic partners	
Do not go alone	
Expensive compared to inorganic fertilizers	

Source: Prepared by the authors.

a very solid foundation and helps identify international opportunities, as well as establish credibility, provide access to critical resources, as well as knowledge, and lead strategic and cooperative partnerships (McDougall, 2000).

Challenges and Opportunities

GIB in the field of organic fertilizers is contributing with an input of fundamental importance in primary agricultural activity. The rehabilitation of organic fertilizer-producing plants in the country is not economically viable in the short term because the reactivation of activities in the plants takes time. Besides, it is not common to find these kinds of producer units in urban areas.

It is necessary to invest more resources in the qualification and to reduce the import of organic inputs and other ingredients necessary in the production of fertilizers. The importation of chemicals as well as organic fertilizers will continue since the national production of 2020 only covers 33.7% of the total demand, and the remaining 66.3% is brought from the foreign market.

Conclusions

When analysing the GIB and eco-efficient company through the comprehensive method of entrepreneurship, it is observed that just as the company has some advantages in the product also with many threats that are the large companies that are already positioned and that also compete with competitive prices. It is also determined that although the GIB has a wide network that is using the best way possible, it is necessary to increase the international network of contacts so that the business can grow quickly.

Although BIO-FOM has been positioned in the local, regional and national markets for a short time and has grown steadily, it has the challenge to enter the international markets in a very competitive position. The company Mary's poop has a competitive advantage that differentiates it from other firms, and that is that its product is innovative and ecological, as well as its processes are carried out in an ecological way which makes it an eco-efficient company. Likewise, due to the culture of planting in Mexico, this company faces problems to become highly competitive in the country since most farmers prefer chemical fertilizers and thus produce more in less time than caring for the land and having a product of a higher quality with organic fertilizers.

It is necessary to implement a culture of conscience in Mexico when we talk about caring for the land since the use of organic fertilizers would not only help the environment but also people; this is because the foods that are consumed will have better nutrients and they would not be contaminated with pesticides. Today some people started to make their gardens at home, so they grow their food; this is a good technique because they can be sure that the consumption of food is 100 per cent organic. This proves the importance of the topic for green innovation in urban areas.

References

Acs, Z. J., & Audretsch, D. B. (2003). *Innovation and technological change*. In Z. J. Acs & D. B. Audretsch (Eds.), *Handbook of entrepreneurship research* (pp. 55–79). Kluwer Academic Publishers.

Ajzen, I. (1991). The theory of planned behavior. *Organizational Behavior and Human Decision Processes*, *50*(2), 1–63.

Aragon-Sanchez, A., Baixauli-Soler, S., & Carrasco-Hernandez, A. J. (2017). A missing link: The behavioral mediators between resources and entrepreneurial intentions. *International Journal of Entrepreneurial Behavior & Research*, *23*(5), 752–768. https://doi.org/10.1108/IJEBR-06-2016-0172

Armitage, C. J. & Conner, M. (2001). Efficacy of the theory of planned behavior: A meta-analytic review. *British Journal of Social Psychology*, *40*(4), 471–499.

Baptista, R. & Preto, M. T. (2010). Long-term effects of new firm formation by type of start-up. *International Journal of Entrepreneurship and Small Business*, *11*(11), 382–402.

Barney, J. (2001). Is the resource-based "view" a useful perspective for strategic management research? Academy of Management Review, *26*(1).

Cantú, P. (2008). Desarrollo sustentable: Conceptos y reflexiones. *Universidad Autónoma de Nuevo León*. Tendencias Científicas. ISBN: 978-607-433-117-2.

Carree, M. A., & Thurik, A. R. (2010). The impact of entrepreneurship on economic growth. In J. Zoltan & D. Audretsch (Eds.), *Handbook of entrepreneurship research*. Springer.

Castro, A. (Julio de 1998). *Hacia el desarrollo sostenible y la ecoeficiencia: Integración de las normas ISO 9000, ISO 14000 e ISO 18000 diseño de un sistema de gestión ambiental certificable. (Tesis de maestría inédita)*. Obtenido de Facultad Regional de Buenos Aires. http://posgrado.frba.utn.edu.ar/investigacion/tesis/MIA-1998-Castro.pdf

Cedillo, G. G. (2021). Interview to a specialist of FIRA conducted on January 13, 2021.

Consejo Mundial Empresarial para el Desarrollo Sostenible. (1991). Ecoeficiencia y sustentabilidad. *Ciencia UANL, 35*.

Evans, D. S., & Leighton, L. S. (1989). Some empirical aspects of entrepreneurship. *American Economic Review, 79*, 519–535.

FIRA (2020, Diciembre 2020). Reporte trimestral de fertilizantes, Dirección de Investigación y Evaluación Económica del Fideicomisos Instituidos en relación con la Agricultura.

Holmes, T. J., & Schmitz, J. A. Jr. (1990). A theory of entrepreneurship and its application to the study of business transfers. *Journal of Political Economy, 98*, 265–294.

Huyghebaert, N. (2004). *Comportamiento estratégico de los titulares en los mercados financieros y la salida de los arranques empresariales*.

Krueger, N. F., Reilly, M. D., & Carsrud, A. L. (2000). Competing models of entrepreneurial intentions. *Journal of Business Venturing, 15*(5–6), 411–432.

Lin, Z., Manser, M. E., & Picot, G. (1998). The role of self-employment in job creation in Canada and the U.S., OECD-CERF-CILN international conference on self-employment, Burlington, Ontario, Canada.

McDougall, P. (2000). *International entrepreneurship*. AMJ.

Ortega-Anderezand, D., & Lai. E. (2017, February). The main factors influencing entrepreneurial activities and intentions within a country – A case study of Spain. *International Journal of Innovation, Management and Technology, 8*(1), 32–37.

Paço, F. D., Arminda, M., Ferreira, J. M., Raposo, M., Gouveia, R., & Dinis, A. (2011). Behaviours and entrepreneurial intention: Empirical findings about secondary students. *Journal of International Entrepreneurship, 9*(1), 20–38.

Peng, M. (2012). *Global strategy*. Thomson South-Western.

Pfeiffer, F., & Reize, F. (2000). Business start-ups by the unemployed – An econometric analysis based on firm data. *Labour Economics, 7*(5), 629–663.

Porter, M. (1980). *Competitive strategy*. Free Press.

Salas, E. B. (2020, September 25). *Volumen de producción mensual de fertilizantes fosfatados en México de enero de 2019 a julio de 2020 (en toneladas métricas)*. Statista. Retrieved 2 July 2021 from https://es.statista.com/estadisticas/594554/volumen-de-produccion-de-fertilizantes-superfosfatados-en-mexico/

Shapero, A., & Sokol, L. (1982). The social dimensions of entrepreneurship. In C. A. Kent, D. L. Sexton, & K. H. Vesper (Eds.), *Encyclopedia of entrepreneurship* (pp. 72–90). Prentice Hall.

Shook, C. L., Priem, R. L., & McGee, J. E. (2003). Venture creation and the enterprising individual: A review and synthesis. *Journal of Management, 29*(3), 379–399.

Souitaris, V., Zerbinatib, S., & Al-Lahamc, A. (2007). Do entrepreneurship programmes raise entrepreneurial intention of science and engineering students? The effect of learning, inspiration and resources. *Journal of Business Venturing, 22*(4), 566–591.

Van Stel, A., & Suddle, K. (2007). The impact of new firm formation on regional development in the Netherlands. *Small Business Economics, 30*, 31–47.

Veciana, J. M., Aponte, M., & Urbano, D. (2005). University students' attitudes towards entrepreneurship: A two countries comparison. *The International Entrepreneurship and Management Journal, 1*(2), 165–182.

Wennekers, A. R. M., & Thurik, A. R. (1999). Linking entrepreneurship and economic growth. *Small Business Economics, 13*, 27–55.

World Business Council for Sustainable Development (1991). *The building blocks of transformation*, www.https://www.wbcsd.org/

Chapter 7

Are the Rich Countries in Terms of Aggregate and Agricultural Activities Responsible for Food Waste? Insights from Some Selected Countries

Buddhadev De and Ramesh Chandra Das

Department of Economics, Vidyasagar University, India

Abstract

Food waste is a challenging and vital issue in the world today. Despite the fact that agricultural technological advancements have effectively improved food production, one-third of worldwide produced foods are being lost or wasted. The fact that large volumes of food are produced but not consumed by humans has serious environmental, social and economic consequences. However the issue of food waste has drawn attention in the last 10 years both in academic and social fields. Under the background, this study aims to examine whether the countries having high agriculture as well as per capita incomes are responsible for food waste or not. The study uses the data on food waste out of export and import channels across the world's selected 20 countries having about 70 per cent contribution to food waste. Using the period 1971–2019, the study finds that food waste is associated with agriculture as well as per capita incomes of the developing and developed countries, positive for majorly developing countries and negative for majorly developed countries. The time series analysis involving food waste, per capita agriculture income and per capita income as the indicators, the study finds that there are long-run relations; among them across many countries implying countries having high incomes are responsible for food waste in most cases due to their food preference, although having good infrastructural facilities for preservation of foods.

Keywords: Food waste; agricultural activity; per capita GDP; food export; food import; developed countries; developing countries

Sustainable Agricultural Practices: Economic and Environmental Implications, 93–114

doi:10.1108/978-1-83608-336-820241007

Introduction

According to the World Bank and Food and Agricultural Organization (FAO), agriculture sector contributes around 4% of global GDP and in some developing countries it accounts for more than 25 per cent of GDP. Agricultural development is one of the most powerful tools to end extreme poverty, leads to reduction in food insecurity, boost share prosperity and feed a projected 9.7 billion people by 2050. Agriculture driven growth, poverty reduction and food security are at risk due to climate change especially world's food-insecure regions. Agriculture, forestry and land use change are responsible for about 25 percent of greenhouse gas emissions. Besides, the sector faces losses and wastes due to improper harvesting, supply chain problems, preference bias towards good quality food, poor logistical supports, among others, which leads to a vast part of the population size unstuffed.

In the literature, food loss refers to a decrease in mass (dry matter) or nutritional value (quality) of food that was originally intended for human consumption. These losses are mainly caused by inefficiencies in the food supply chains, such as poor infrastructure and logistics, lack of technology, insufficient skills, knowledge and management capacity of supply chain actors, and lack of access to markets. In addition, natural disasters play a role. On the other hand, food waste refers to food appropriate for human consumption being discarded, whether or not after it is kept beyond its expiry date or left to spoil. Often this is because food has spoiled but it can be for other reasons such as oversupply due to markets or individual consumer shopping/eating habits. Food wastage refers to any food lost by deterioration or waste. Thus, the term 'wastage' encompasses both food loss and food waste. Throughout the whole process of food supply chain *food losses* occur at production, post-harvest and processing stages. On the other hand, food losses at the end of the food chain (retail and final consumption are called food waste (Parfitt et al., 2010). The world population of 21st century, which is much larger now and this population burden, will fall largely 'A livelihood comprises people, their capabilities and their means of living, including food, income and assets' (Chambers & Conway, 1991).

Agricultural processes are self-employed activities throughout the world especially in case of developing countries. People, who are self-employed in agricultural activities and are more skilled, are able to raise farm earnings within the same farm area. These complementarities result from the fact that more skilled farmers are able to make prompt and efficient decisions regarding crop selection, input substitution and output market targeting, which raises farm earnings within the same farm area. This, in turn, leads to increased access to diversified and nutritious food (apart from traditional farm-grown food) and to a more sustainable means of subsistence. They will experience improved personal and child health outcomes, more home comforts, a sense of food security, and most importantly, gender mainstreaming in this situation. The gendering of livelihood arises from the fact that, once livelihood sustainability is achieved (or is thought to be achieved in the near future), women have more options for leisure, education, work and social status security.

Agricultural income of developing countries has far-reaching implications for agricultural growth and poverty alleviation. It also contributes to alleviate rural poverty. Ensuring high agricultural income enlightens the livelihood of the farmers and the promotion of the livelihood empowered the whole family members in nutrition, education and also in health facilities (Singh, 2013).

In the world, family farming is a major form of agricultural production. To pursuing different agricultural and pastoral management activities, they cover a wide range of relatively small structures, production, productivity or level of capitalization in accordance with their various social agroecological as well as economic conditions using limited land holdings. They use their production for both own consumption and sale, mostly relying on family labour but also employing the same poor or less fortunate neighbours (Rodrigues-Fortes, 2020). The farm owned by a family with larger economic size can develop with the market where family members engage in agricultural activity and influences with their decision. Whereas in case of small farms agricultural income has significant share in total income but show the tendency for diversification of agriculture activities (Chmieliński et al., 2023). People can support family farmers and small businesses in their community by buying local produce (FAO, 2020).

It is critical to reduce poverty by increasing income but improving upon agriculture incomes have the greatest impact. An increase in agriculture income requires an increase in production, which can lower down income inequality as well as consumption disparities (Oduro & Osei-Akoto, 2008).

Number of countries is increasing for both significant importers and exporters of food. Since that self-sufficient countries in food production have an excess food producer they also import some food products. On the other hand, less-sufficient food producing countries also export some food products. Through the technological advancements the shipment of food products is easy from one side to other side of the world; but within a few days during or after transportation storage it may get spoiled or unsafe when it reaches the consumer.

International trade on food items play a crucial role in (i) availability of food in sufficient quantities and appropriate quality throughout import, (ii) access of nutritious diet, (iii) utilization of food to reach a state of nutritional well-being and (iv) stability of access to food all time without risks of sudden shocks. In global food markets high-income economies meet about 60 percent of their food needs whereas the low-income countries import only about 14 percent of their food consumption. That is why richer and large countries can better leverage comparative advantage where low-income countries face a high trade costs (WTO Committee on Agriculture, 2023).

In food supply chain producers are at the primary level. Food losses occur even before food reaches the shelves, on farms. Another reason of food getting wasted in field is the imperfect appearance of fruits and vegetables in USA. The retailers and consumers have demanded 'pretty fruit' which started a process of 'selective harvesting' where at most 30% of crops which do not meet quality or appearance are intentionally left behind or rot (Solis, 2019). According to FAO, it is roughly one-third throughout the world that 1.3 billion tonnes of food produced for human consumption get wasted or lost every year which is worth US$680 billion in industrialized countries and US$310 billion in developing countries. Therefore, it appears that developed countries waste more food in processing, retail and consumption stages in most cases. On the other hand, developing countries waste food during production and harvesting phases.

Literature Review

In 2011, FAO in its first report pointed out that one-third of all food produced for human consumption in the world is lost or wasted. This food wastage represents a missed opportunity to improve global food security and to mitigate environmental impacts generated by agriculture. Alexandratos and Bruinsma (2012) find that by 2050, food production will need to be 60% higher than in 2005/2007, if production is to meet demand of the increasing world population. Making better use of food already available with the current level of production would help to meet future demand with a lower increase in agricultural production. According to (FAO 2011, 2009), 10% of rich countries' greenhouse gas emissions come from growing food that is never eaten. Also the irrigation water used globally to grow food that is wasted would be enough for the domestic needs (at 200 L per person per day) of 9 billion people – the number expected on the planet by 2050.

Michael et al. (2003) provide a framework for analysing the (positive and negative) impact of irrigation on poverty and approaches that irrigation is most likely to have a positive impact on poor basically small farmers, landless laborers and urban poor. Weber and Khademian (2008) depicted the negative impact of irrigated agriculture on environment sustainability as a 'wicked problem' since it is cross-cutting across multiple policy sectors and levels, unstructured due to uncertainty and disagreement about its causes and solutions and relentlessly with never-ending issues at stake. Cervantes-Godoy and Dewbre (2010) in their work show that economic growth generally was an important contributor to poverty reduction, the sector mix of growth mattered substantially. Majority of the developing country were put a great importance and more investment on the agricultural sector growth for poverty reduction.

The Global Food Security Programme Report (2013) depicts that the global population exceeded seven billion people during 2011 and is predicted to reach 9.3 billion by 2050. Alongside more mouths to feed, increasing economic development allows people to consume more, leading to a projected increased food demand of 50–70% by mid-century. Reducing global food waste will have a significant part to play in increasing the availability of food in the future. Tielens and Candel (2014) depict about the UN 'Zero lost or waste of food'. If food wastage is reduced, less land, water, inputs and energy are needed, and less greenhouse gas emitted. These natural resources could be used to increase food production or affect the food system in other ways. Attenborough (2012) suggests that hunger cannot be alleviated nor famines avoided only by making the food system environmentally sound. But neither can food be produced perpetually by eroding its ecological foundation. So a secure ecological foundation is a necessary condition for secure food system. Food is not only a form of energy but also a consumer of fossil energy in its production, transportation and preparation. The last 50 years have seen increased agricultural productivity, thanks to the adoption of new technologies and inputs, which are largely based on fossil fuels. Here the energy required to produce food is wasted or lost due to food loss or wasted food (Cuellar & Webber, 2010).

Global trade contribution of South Asia has remained at 1% and generates only 2% of the global income, yet supports 22% of the world's population and 44% of the world's poor. Over 75% of the population depends directly or indirectly on agriculture for their livelihoods (Gunasena, 2003). Post-harvest losses refer to food damage and degradation during different post-harvest handling and processing stages from the beginning and completion and harvesting to the moment of final consumption. They include harvesting losses, such as when grain is threshed, winnowed and dried (Sheahan & Barret, 2017). Griffin et al. (2009) described that a reduction or a better use of wastage throughout the food chain would contribute to a more efficient use of natural resources and nutrients, and through this it would decrease the environmental and climate impact of food production and consumption.

Research Gaps

To date, no study has analysed the impact of per capita GDP (PCGDP) and per capita agricultural GDP (PCAGDP) on global food waste in per capita terms. The present study wants to examine that whether agricultural-dependent countries are wasting more food than agriculture-dependent countries. They also want to examine further whether the high-income countries waste more food than the low-income countries?

Objectives of the Study

1. To analyse the trends of per capita food waste (PCFW) due to export and import, PCGDP and PCAGDP for world's 40 developed and developing countries of the world for the period 1971–2019.
2. To find the long-run relationships between PCFW, PCGDP and PCAGDP for the listed countries.

Data and Methodology

Data

Here the study first collects country-wise GDP, agricultural GDP, food waste due to export, food waste due to import and population data from the databases of the World Bank, Food and Agricultural Organization, and Our World in Data for the year 1971–2019. Then we calculate PCGDP and PCAGDP from GDP and agricultural GDP by dividing the population for the mentioned year. To derive PCFW the study first sums up food waste due to export and food waste due to import data for the same country according to year. Then by dividing it by the value of population it calculates PCFW data.

The study first uses descriptive statistics to compare cross-country food waste, PCAGDP and PCGDP in the selected 40 (20 developed and 20 developing) countries of the world for the period 1971–2019.

In the second step, the study makes correlation of food waste with PCAGDP and PCGDP to find whether the countries with high PCAGDP and high PCGDP are associated with higher PCFW.

Time series econometric analysis for individual countries is done in the sequence of first testing of stationarity by ADF test techniques. Second, the Johansen cointegration technique is followed for examining long-run relation in a vector auto regression (VAR) set up. Third, the error correction mechanism is applied for examining short-run dynamics around the long-run relations. Finally, the test of causality/Block Exogeneity is done for investigating short-run interplays among the selected indicators, PCGDP, PCAGDP and PCFW.

Results and Discussion

Descriptive Statistics

We have the data for PCAGDP, PCGDP and PCFW from 1971 to 2019. We divide this time span into two time period, i.e., the first one from 1971 to 1995 and the last one from 1996 to 2019 for the purpose of pre- and post-WTO analysis. Fig. 7.1 (Panels A to F) displays the trends of the selected indicators for both the groups of developed and developing countries. There are several observations from the figure. Firstly, in case of developed countries, it is observed that agricultural dependence is decreasing after 1995 and PCGDP is increased for the same time period. Along with this their PCFW in supply chain has also increased. On the other hand, agricultural dependency as measured by the PCAGDP, a rough proxy, of developing countries is also decreasing after 1995 and PCGDP have slow increasing trends and over the time the trends of PCFW are also increasing.

Average PCAGDP for the entire period (1971–2019) is \$460 in developing countries, and for the developed countries, it is \$700, whereas average PCGDP in the developing countries is \$4,650 for the entire period (1971–2019), and for the developed countries, it is \$33,625. So, from the above observation we can say that the developed countries are far away than the developing countries in terms of PCAGDP as well as PCGDP.

Comparison Between PCAGDP and PCFW

From Panel A we can see for the developed countries that the PCAGDP has downward trends for countries like Austria, Cyprus, Denmark, Finland, France, Germany, Greece, Italy, Mauritius, Netherlands, Portugal, Sweden, Switzerland and United Kingdom. But the trends of PCFW are upward rising for the same countries in Panel E. It means that developed countries have decreased their PCAGDP throughout years while their PCFW have increased at the same time. On the other hand, from Panel B, we can see for developing countries that PCAGDP have downward trends for Brazil, Colombia, Costa Rica, Guatemala, South Africa and Turkey. Here also we can see an upward trend in eight countries. The trend of PCFW is upward as shown in Panel F.

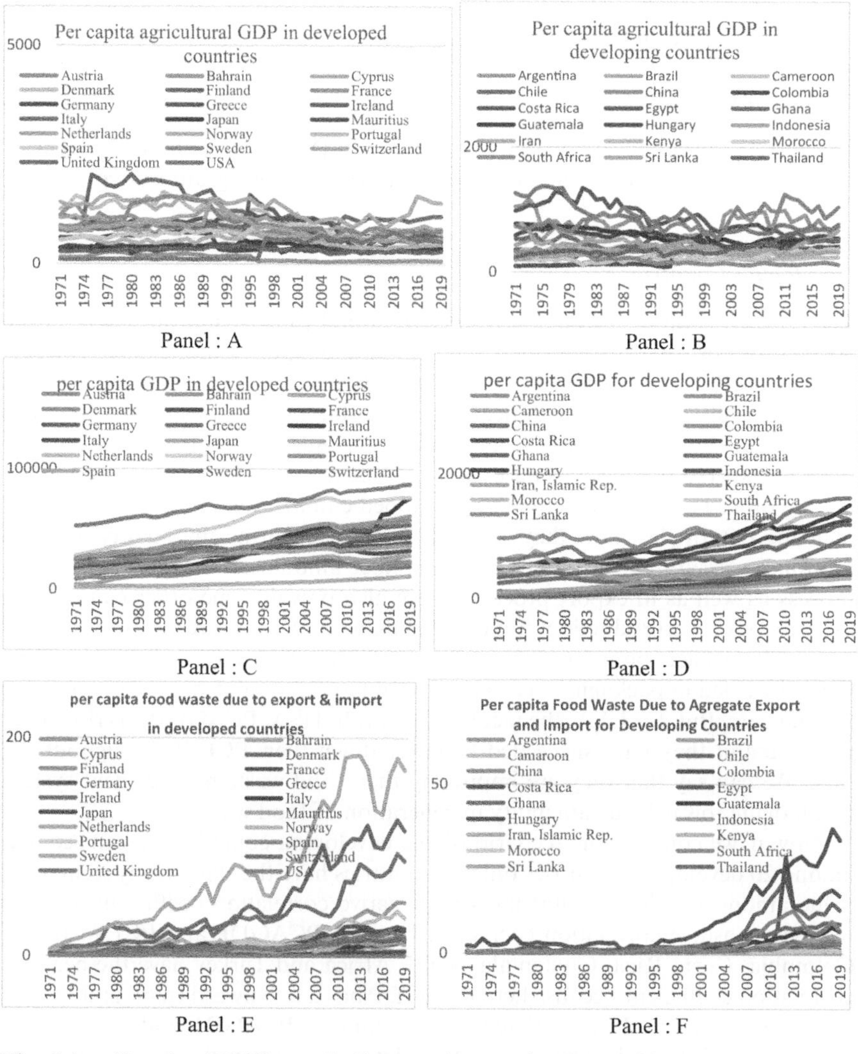

Fig. 7.1. Trends of Different Indicators Across the Countries. *Source*: Drawn by the authors.

Comparison Between PCGDP and PCFW

From the Panel C we can see for the developed countries that the PCGDP has upward trends for the countries Austria, Cyprus, Denmark, Finland, France, Germany, Greece, Italy, Japan, Mauritius, Netherlands, Portugal, Spain, Sweden, Switzerland and United Kingdom. But, on the other hand, if we see the trend for PCFW in Panel E we see that there are upward trends for the same countries. Now if we consider about the developing countries, we can see in Panel D that among 20 developing countries 17 countries PCGDP have upward trend. We have mentioned earlier in Panel F that the trend of PCFW is also upward. Now the study will analyse the sign of association with the help of correlation between PCAGDP and PCGDP with PCFW.

Correlation Results

The correlation results are presented in Panel A of Table 7.1. It shows the signs of correlation coefficients in pairs of the indicators in the two groups of countries.

Correlation Between PCAGDP and PCFW for Developed and Developing Countries

The countries whose correlation results are found to be significant either in the negative or positive directions only their results are reported in the table. The correlation coefficients between PCAGDP and PCFW for the developed countries are statistically significant with negative values for 15 countries out of 20. Since the developed countries are not agricultural-dependent, their per capita agricultural output (AO) is decreasing over time; but their food waste are increasing. On the other hand as the people of developed countries improve their standard of living day by day, now they are in the aim of ensuring the quality of food. For quality assurance they used to waste a huge amount of food through import process as well as in export process. There also prevail some other causes such as overstock, overproduction, improper demand and sales forecast, unprofitable prices of produce, poor food management skills/poor food purchase planning, food discarded in packaging because of misinterpretation of best before and use by dates, etc. But, for USA, the fact is different. In USA, PCAGDP has risen over the time and they have succeeded to lower down their PCFW throughout the time. We can say that they are improving their food waste management system and moving towards sustainability in generation of food waste.

Now if we consider the scenario of the developing countries we can see that among 20 developing countries only 14 countries have the statistically significant value among which six countries have negative correlation coefficient and the rest have positive correlation coefficient between PCAGDP and PCFW. Brazil, Colombia, Costa Rica, Guatemala, South Africa and Turkey have the negative correlation coefficient because here PCAGDP has a downward trend and PCFW has a positive trend. In case of remaining countries PCAGDP and PCFW both have the positive trends. The causes of food waste in the developing countries are not the supply chain problems like the developed countries but they suffer from infrastructural and logistical problems which hinder them from proper storage and transport of the food products from and to the countries. But Iran has the negative trend for those two, i.e., PCFW is falling over time with PCAGDP.

Correlation Between PCGDP and PCFW for Developed and Developing Countries

The correlation coefficients between PCGDP and PCFW for the developed countries are statistically significant with positive values for 18 countries out of 20 (Table 7.1, Panel A). The countries are Austria, Cyprus, Denmark, Finland,

Table 7.1. Correlation and Stationarity Test Results.

Panel A: Signs of Correlation Coefficients in Pairs of the Indicators

Correlation Between	Sign of Association	Developed	Signs of the trends		Developing	Signs of the trends		Remarks
PCAGDP and PCFW	+	00	+	PCAGDP	08	+	PCAGDP	Iran has two '−ve'
			+	PCFW		+	PCFW	
	−	15	+	PCFW	06	+	PCFW	
			−	PCAGDP		−	PCAGDP	
PCGDP and PCFW	+	18	+	PCGDP	17	+	PCGDP	
			+	PCFW		+	PCFW	
	−	01 (USA)	+	PCGDP	00	+	PCGDP	
			−	PCFW		−	PCFW	

Panel B: Stationarity Test Results of the Selected Countries

Country	PCFW	PCAGDP	PCGDP	Country	PCFW	PCAGDP	PCGDP
Argentina	−6.72 (0.00)	−8.26 (0.00)	−5.95 (0.00)	Austria*	−6.40 (0.00)	−4.75 (0.00)	−8.67 (0.00)
Brazil	−5.99 (0.00)	−7.41 (0.00)	−4.96 (0.002)	Bahrain	−7.69 (0.00)	−5.90 (0.00)	−5.15 (0.00)
Cameroon	−7.35 (0.00)	−7.61 (0.00)	−4.13 (0.00)	Cyprus	−8.01 (0.00)	−3.71 (0.00)	−9.05 (0.00)
Chile	−3.64 (0.00)	−6.44 (0.00)	−4.32 (0.00)	Denmark	−10.53 (0.00)	−11.1 (0.00)	−9.41 (0.00)
China*	−6.55 (0.00)	−4.60 (0.00)	0.003 (0.95)	Finland	−8.49 (0.00)	−7.16 (0.00)	−9.41 (0.00)
Colombia	−6.26 (0.00)	−5.54 (0.00)	−3.74 (0.00)	France	−6.87 (0.00)	−5.09 (0.00)	−8.34 (0.00)

(*Continued*)

Table 7.1. (*Continued*)

Panel B: Stationarity Test Results of the Selected Countries							
Costa Rica	−3.68 (0.00)	−7.42 (0.00)	−4.29 (0.00)	Germany	−6.19 (0.00)	−6.76 (0.00)	−7.72 (0.00)
Egypt	−7.27 (0.00)	−7.98 (0.00)	−4.33 (0.00)	Greece	−8.06 (0.00)	−8.57 (0.00)	−3.56 (0.00)
Ghana	−8.45 (0.00)	−6.75 (0.00)	−3.95 (0.00)	Ireland	−6.82 (0.00)	−7.72 (0.00)	−3.82 (0.00)
Guatemala	−6.49 (0.00)	−6.32 (0.00)	−2.99 (0.00)	Italy	−7.33 (0.00)	−8.63 (0.00)	−4.50 (0.00)
Hungary	−6.82 (0.00)	−6.80 (0.00)	−3.49 (0.00)	Japan	−6.33 (0.00)	−9.81 (0.00)	−5.39 (0.00)
Indonesia	−7.95 (0.00)	−6.58 (0.00)	−4.10 (0.00)	Mauritius*	−6.82 (0.00)	−11.10 (0.00)	−9.41 (0.00)
Iran	−8.39 (0.00)	−6.32 (0.00)	−4.63 (0.00)	Netherlands	−5.76 (0.00)	−7.66 (0.00)	−4.38 (0.00)
Kenya*	3.39 (0.00)	−8.22 (0.00)	−6.10 (0.00)	Norway	−8.38 (0.00)	−6.40 (0.00)	−3.40 (0.01)
Morocco	−10.84 (0.00)	−7.17 (0.00)	−7.49 (0.00)	Portugal	5.75 (0.00)	−7.74 (0.00)	−4.04 (0.00)
South Africa	−7.74 (0.00)	−7.07 (0.00)	−4.06 (0.002)	Spain*	−7.71 (0.00)	−8.02 (0.00)	−6.83 (0.00)
Sri Lanka	−6.37 (0.00)	−7.00 (0.00)	−2.98 (0.04)	Sweden	−5.65 (0.00)	−6.38 (0.00)	−5.36 (0.00)
Thailand	−6.66 (0.00)	−6.02 (0.00)	−4.92 (0.00)	Switzerland	−7.33 (0.00)	−8.63 (0.00)	−4.50 (0.00)
Turkey	−7.22 (0.00)	−6.39 (0.000	−5.85 (0.00)	UK	−2.97 (0.04)	−7.26 (0.00)	−4.80 (0.00)
Uruguay	−4.51 (0.00)	−7.55 (0.00)	−3.49 (0.00)	USA	−7.13 (0.00)	−7.21 (0.00)	−4.94 (0.00)

'*' – Stationary at second difference.

Source: Authors' computations.

France, Germany, Greece, Italy, Japan, Mauritius, Netherlands, Portugal, Spain, Sweden, Switzerland and United Kingdom. But, on the other hand, if we see the trend for PCFW, we see that there are upward trends for the same list of the countries. It means that developed countries have increased their PCGDP throughout years while their PCFW have increased at the same time. But for USA the fact is different. In USA PCGDP has risen over the time and they have succeeded to lower down their PCFW throughout the time. We can say that they are improving their food waste management system and moving towards sustainability in generation of food waste. Now if we consider about the developing countries we can see in Panel D that among 20 developing countries 17 countries have the statistically significant values with positive correlation coefficient between PCGDP and PCFW. The countries' PCGDP has upward trend and countries' positive trend for PCFW makes the correlation coefficient positive. Hence the correlation results mean that the rich countries are making more food waste due to their preference bias and the relatively poor countries are making less food waste.

Stationarity Test Results

We test the nature of the time series of the selected indicators through ADF test of unit root problem. The results are given in Panel B of Table 7.1.

From Panel B of the table we get that out of all the 40 countries the major countries' indicators, i.e., PCFW, PCAGDP and PCGDP are integrated of order 1. Bahrain, Cyprus, Denmark, Finland, France, Germany, Greece, Ireland, Italy, Japan, Netherlands, Norway, Portugal, Sweden, Switzerland, United Kingdom and USA are the developed countries in the list, and the list of the developing countries are Argentina, Brazil, South Africa, Sri Lanka, Cameroon, Chile, Colombia, Costa Rica, Egypt and Ghana. Besides, there are a low number of countries, namely, Austria, Mauritius, Spain, China and Kenya where the indicators are integrated of order 2.

There is the possibility of long-run association between PCAGDP, PCGDP and PCFW in the listed 40 countries who maintain I(1) or I(2) in their indicators. The study thus conducted Johansen cointegration test to examine long-run association between PCAGDP, PCGDP and PCFW.

Cointegration Test Results

Table 7.2 presents the Johansen cointegration results of the listed countries involving three time series variables.

From the table it is observed that among 20 developed countries the long-run relationship between the selected variables is found for seven countries, namely, Austria, Bahrain, Denmark, France, Ireland, Mauritius and United Kingdom. On the other hand, for 20 developing countries, the long-run relationship between the selected variables is found for 11 countries such as Argentina, China, Egypt, Ghana, Indonesia, Iran, Kenya, Morocco, South Africa, Sri Lanka and Uruguay. Now we proceed for examining the short-run dynamics around the long-run relations through VECM, and the results are shown in Table 7.3.

Table 7.2. Cointegration Test Results for Developing and Developed Countries.

Developing Country	Hypothesized No. of CE(s)	Trace Statistics (Prob)	Developed Country	Hypothesized No. of CE(s)	Trace Statistics (Prob)
Argentina#	None*	33.15086 (0.02)	Austria#	None*	29.97011 (0.04)
	At most 1	10.71443 (0.23)		At most 1	14.22815 (0.07)
	At most 2	1.101638 (0.29)		At most 2	2.900682 (0.08)
Brazil	None*	24.58876 (0.17)	Bahrain#	None*	29.76934 (0.05)
	At most 1	6.888526 (0.59)		At most 1	8.712701 (0.39)
	At most 2	0.432991 (0.51)		At most 2	1.456846 (0.22)
Cameroon	None*	12.70074 (0.90)	Cyprus	None*	20.17002 (0.41)
	At most 1	3.867099 (0.91)		At most 1	6.094569 (0.68)
	At most 2	0.382823 (0.53)		At most 2	0.002974 (0.95)
Chile	None*	15.12918 (0.77)	Denmark#	None*	29.44579 (0.05)
	At most 1	7.38915 (0.53)		At most 1	14.30605 (0.07)
	At most 2	0.264658 (0.60)		At most 2	0.701366 (0.40)
China#	None*	33.98256 (0.01)	Finland	None*	20.61458 (0.38)
	At most 1*	7.543751 (0.51)		At most 1*	6.683774 (0.61)
	At most 2	0.271287 (0.60)		At most 2	0.395677 (0.52)
Colombia	None*	18.34815 (0.54)	France#	None*	30.85186 (0.03)
	At most 1	5.752956 (0.72)		At most 1	6.781968 (0.60)
	At most 2	0.137449 (0.71)		At most 2	2.559159 (0.10)
Costa Rica	None*	26.19686 (0.12)	Germany	None *	15.45208 (0.75)
	At most 1*	8.645912 (0.39)		At most 1*	5.429439 (0.76)
	At most 2*	0.845009 (0.35)		At most 2*	0.003721 (0.95)

Egypt#	None*	29.68047 (0.05)	Greece	None*	26.00179 (0.12)
	At most 1	8.438625 (0.42)		At most 1	9.963475 (0.28)
	At most 2	1.76977 (0.18)		At most 2	1.699694 (0.19)
Ghana#	None*	34.71988 (0.01)	Ireland#	None*	32.68074 (0.02)
	At most 1*	14.10342 (0.08)		At most 1*	7.220623 (0.55)
	At most 2*	0.651124 (0.41)		At most 2*	2.022958 (0.15)
Guatemala	None*	20.20344 (0.40)	Italy	None*	26.17102 (0.12)
	At most 1*	8.052148 (0.45)		At most 1*	11.85763 (0.16)
	At most 2	0.117234 (0.73)		At most 2	0.387353 (0.53)
Hungary	None	25.43988 (0.14)	Japan	None	15.10038 (0.77)
	At most 1	11.22839 (0.19)		At most 1	5.149734 (0.79)
	At most 2	1.235 (0.26)		At most 2	0.351373 (0.55)
Indonesia#	None*	54.07168 (0.00)	Mauritius#	None*	36.36587 (0.00)
	At most 1	13.75169 (0.09)		At most 1	16.06514 (0.04)
	At most 2*	2.689151 (0.10)		At most 2*	5.251518 (0.02)
Iran#	None*	32.89871 (0.02)	Netherland	None*	18.30852 (0.54)
	At most 1*	13.81314 (0.08)		At most 1*	5.042847 (0.80)
	At most 2	2.158637 (0.14)		At most 2	0.003855 (0.94)
Kenya#	None*	35.5454 (0.00)	Norway	None*	22.01195 (0.29)
	At most 1*	4.782072 (0.83)		At most 1*	6.722553 (0.61)
	At most 2	0.010243 (0.91)		At most 2	3.096501 (0.07)

(*Continued*)

Table 7.2. (*Continued*)

Developing Country	Hypothesized No. of CE(s)	Trace Statistics (Prob)	Developed Country	Hypothesized No. of CE(s)	Trace Statistics (Prob)
Morocco#	None*	33.60068 (0.01)	Portugal	None*	17.95064 (0.56)
	At most 1*	15.29996 (0.05)		At most 1*	5.821251 (0.71)
	At most 2	3.735809 (0.05)		At most 2	0.012515 (0.91)
South Africa#	None	36.44839 (0.00)	Spain#	None	41.19585 (0.00)
	At most 1	11.70543 (0.17)		At most 1	12.53054 (0.13)
	At most 2*	0.82939 (0.36)		At most 2*	0.945631 (0.33)
Sri Lanka#	None	40.67901 (0.00)	Sweden	None	20.42665 (0.39)
	At most 1	12.16934 (0.14)		At most 1	6.349041 (0.65)
	At most 2*	2.368947 (0.12)		At most 2*	0.21989 (0.63)
Thailand	None	23.13195 (0.23)	Switzerland	None	14.35761 (0.81)
	At most 1	4.587783 (0.85)		At most 1	6.923437 (0.58)
	At most 2*	0.337869 (0.56)		At most 2*	0.000797 (0.97)
Turkey	None	25.30117 (0.15)	UK#	None	24.28439 (0.18)
	At most 1	5.790206 (0.72)		At most 1	9.252309 (0.34)
	At most 2*	0.24741 (0.61)		At most 2*	3.983534 (0.04)
Uruguay#	None	30.8853 (0.03)	USA	None	14.35761 (0.81)
	At most 1	10.86193 (0.22)		At most 1	6.923437 (0.58)
	At most 2*	4.758172 (0.02)		At most 2*	0.000797 (0.97)

Source: Authors' computations.

Note: '#' – Cointegration is found.

Table 7.3. VECM Results for the Developed and Developing Countries.

Developing Country	Dependent Variable	Coefficient of ECT (Prob.)	Remarks	Developed Country	Dependent Variable	Coefficient of ECT (Prob.)	Remarks
Argentina	PCFW	−0.56 (0.00)	PCAGDP, PCGDP → PCFW	Austria	PCFW	−0.00 (0.62)	No LR causality
	PCAGDP	−31.87 (0.19)	No LR causality		PCAGDP	0.79 (0.00)	No LR causality
	PCGDP	27.06 (0.79)	No LR causality		PCGDP	3.62 (0.06)	No LR causality
China	PCFW	0.13 (0.00)	No LR causality	Bahrain	PCFW	0.01 (0.01)	No LR causality
	PCAGDP	35.26 (0.00)	No LR causality		PCAGDP	−0.04 (0.76)	No LR causality
	PCGDP	−51.67 (0.09)	No LR causality		PCGDP	−12.62 (0.00)	PCFW, PCAGDP → PCGDP
Egypt	PCFW	−0.01 (0.52)	No LR causality	Denmark	PCFW	−0.03 (0.23)	No LR causality
	PCAGDP	−6.39 (0.00)	PCFW, PCGDP → PCAGDP		PCAGDP	−1.70 (0.00)	PCFW, PCGDP → PCAGDP
	PCGDP	−3.75 (0.17)	No LR causality		PCGDP	0.72 (0.78)	No LR causality
Ghana	PCFW	0.00 (0.89)	No LR causality	France	PCFW	−0.00 (0.93)	No LR causality
	PCAGDP	10.03 (0.00)	No LR causality		PCAGDP	1.86 (0.00)	No LR causality
	PCGDP	16.01 (0.00)	No LR causality		PCGDP	2.13 (0.44)	No LR causality
Indonesia	PCFW	−0.57 (0.00)	PCAGDP, PCGDP → PCFW	Ireland	PCFW	−0.09 (0.29)	No LR causality
	PCAGDP	745.75 (0.00)	No LR causality		PCAGDP	−3.85 (0.16)	No LR causality
	PCGDP	1654.26 (0.00)	No LR causality		PCGDP	106.99 (0.00)	No LR causality
Iran	PCFW	−0.52 (0.03)	PCAGDP, PCGDP → PCFW	Mauritius	PCFW	0.00 (0.90)	No LR causality
	PCAGDP	168.70 (0.01)	No LR causality		PCAGDP	1.57 (0.03)	No LR causality
	PCGDP	−139.73 (0.76)	No LR causality		PCGDP	−2.96 (0.04)	PCFW, PCAGDP → PCGDP

(Continued)

Table 7.3. (*Continued*)

Developing Country	Dependent Variable	Coefficient of ECT (Prob.)	Remarks	Developed Country	Dependent Variable	Coefficient of ECT (Prob.)	Remarks
Kenya	PCFW	0.69 (0.00)	No LR causality	UK	PCFW	−0.01 (0.08)	No LR causality
	PCAGDP	36.25 (0.27)	No LR causality		PCAGDP	−0.57 (0.01)	PCFW, PCGDP → PCAGDP
	PCGDP	−30.42 (0.39)	No LR causality		PCGDP	11.92 (0.04)	No LR causality
Morocco	PCFW	−0.03 (0.74)	No LR causality				
	PCAGDP	−54.81 (0.17)	No LR causality				
	PCGDP	215.81 (0.02)	No LR causality				
South Africa	PCFW	−0.05 (0.00)	PCAGDP, PCGDP → PCFW				
	PCAGDP	−0.58 (0.73)	No LR causality				
	PCGDP	−1.48 (0.85)	No LR causality				
Sri Lanka	PCFW	−0.00 (0.91)	No LR causality				
	PCAGDP	−1.27 (0.00)	PCFW, PCGDP → PCAGDP				
	PCGDP	0.65 (0.45)	No LR causality				
Uruguay	PCFW	−0.35 (0.02)	PCAGDP, PCGDP → PCFW				
	PCAGDP	33.61 (0.06)	No LR causality				
	PCGDP	−2.15 (0.97)	No LR causality				

Source: Authors' computations.

VECM Results

From Table 7.3 it is observed that the errors around the cointegrating relation are found to be negative and statistically significant for the developing countries Argentina, Egypt, Indonesia, Iran, South Africa, Sri Lanka and Uruguay. For the developed countries, it is observed that the errors around the cointegrating relation are found to be negative and statistically significant only for Bahrain, Denmark, Mauritius and UK. This means that the errors are temporary in nature and they move back to the original long-run relation. Thus, the long-run relation is stable for these sets of countries. It is also to say that there are long-run causal influences from PCAGDP and PCGDP to PCFW. This alternatively means agricultural dependence and per capita income levels are the important causes of food waste in the said countries.

Results of Short-Run Causality Through Wald Test (Block exogeneity Test/Granger Causality in VAR)

The short-run causal interplays are analysed using Wald test. The hypotheses are given as

> H_0: PCAGDP and PCGDP simultaneously do not cause PCFW.
> H_1: PCAGDP and PCGDP cause PCFW

The results are given in Table 7.4.

We find that only in case of four developing countries: China, Morocco, Sri Lanka and Uruguay that PCAGDP and PCGDP make cause to PCFW. Further, we find that only in case of two developed countries, Finland and Portugal, that PCAGDP and PCGDP make cause to PCFW. Greece, Sweden and United Kingdom are three countries where PCAGDP make cause PCFW.

Discussion

In agriculture there is no significant difference in PCAGDP for developed and developing countries. The sources of high income in the developed countries are mainly industry and service sector. Transport and storage facility are better in developed countries. So the wastage of food through transport and storage is lesser in the developed countries than the developing countries. In developed countries a significant amount of food is noticed to be wasted in consumption stage. Even a significant amount of food losses also occurred in early of the food supply chain which do not maintain quality assurance (i.e. perfect in shape or appearance). Mainly it is found that preference bias creates voluntary nature of food waste.

If we compare PCGDP for developed and developing countries we can see that PCGDP is much higher for the developed countries. Due to high PCGDP, with their higher income they consider the quality and appearance of food. Only for this reason the PCFW for the developed countries is computed at $14 (for the time period 1971–2019) whereas the developing countries' PCFW is computed at $1.6.

Table 7.4. Results of Short-Run Causality Through Wald Test for the Selected Developed and Developing Countries.

Developing Country	Dependent Variable	Independent Variables	Chi-Square Value (Prob)	Remarks	Developed Country	Chi-Square Value (Prob)	Remarks
Argentina	PCFW	PCAGDP	1.30 (0.52)	–	Austria	1.01 (0.60)	
		PCGDP	0.98 (0.61)	–		0.59 (0.74)	
		All	1.67 (0.61)	–		1.54 (0.81)	
Brazil	PCFW	PCAGDP	0.17 (0.91)	–	Bahrain	3.62 (0.16)	
		PCGDP	0.348 (0.84)	–		0.01 (0.99)	
		All	0.45 (0.97)	–		3.72 (0.44)	
Cameroon	PCFW	PCAGDP	2.59 (0.27)	–	Cyprus	0.56 (0.75)	
		PCGDP	2.68 (0.26)	–		1.71 (0.42)	
		All	3.65 (0.45)	–		1.96 (0.74)	
Chile	PCFW	PCAGDP	0.43 (0.80)	–	Denmark	0.27 (0.87)	
		PCGDP	0.79 (0.67)	–		0.00 (0.99)	
		All	1.05 (0.90)	–		0.27 (0.99)	
China	PCFW	PCAGDP	3.53 (0.17)	–	Finland	1.06 (0.58)	
		PCGDP	14.20 (0.00)	PCGDP → PCFW		4.98 (0.08)	
		All	18.28 (0.00)	PCGDP, PCAGDP → PCFW		8.91 (0.05)	PCAGDP, PCGDP → PCFW
Colombia	PCFW	PCAGDP	0.65 (0.72)	–	France	0.16 (0.91)	
		PCGDP	4.25 (0.11)	–		1.38 (0.49)	
		All	5.53 (0.23)	–		1.90 (0.75)	

Costa Rica	PCFW	PCAGDP	0.84 (0.65)	–	Germany	0.29 (0.86)	
		PCGDP	3.71 (0.15)	–		2.95 (0.22)	
		All	4.23 (0.37)	–		3.13 (0.53)	
Egypt	PCFW	PCAGDP	0.45 (0.79)	–	Greece	1.85 (0.39)	
		PCGDP	0.01 (0.99)	–		6.16 (0.05)	PCGDP → PCFW
		All	0.48 (0.97)	–		6.73 (0.15)	
Ghana	PCFW	PCAGDP	0.89 (0.64)	–	Ireland	1.92 (0.38)	
		PCGDP	0.86 (0.64)	–		1.15 (0.56)	
		All	2.11 (0.71)	–		3.30 (0.50)	
Guatemala	PCFW	PCAGDP	0.24 (0.88)	–	Italy	2.35 (0.30)	
		PCGDP	0.51 (0.77)	–		1.90 (0.38)	
		All	0.81 (0.93)	–		4.30 (0.36)	
Hungary	PCFW	PCAGDP	0.44 (0.80)	–	Japan	0.69 (0.70)	
		PCGDP	1.18 (0.55)	–		0.89 (0.64)	
		All	1.42 (0.55)	–		1.91 (0.75)	
Indonesia	PCFW	PCAGDP	0.98 (0.60)	–	Mauritius	1.93 (0.38)	
		PCGDP	0.14 (0.93)	–		4.11 (0.12)	
		All	1.25 (0.93)	–		4.59 (0.33)	
Iran	PCFW	PCAGDP	1.65 (0.43)	–	Netherlands	1.51 (0.47)	
		PCGDP	0.22 (0.89)	–		0.50 (0.77)	
		All	2.23 (0.69)	–		2.30 (0.67)	
Kenya	PCFW	PCAGDP	1.95 (0.58)	–	Norway	1.25 (0.53)	
		PCGDP	5.72 (0.12)	–		0.61 (0.73)	
		All	7.20 (0.30)	–		1.99 (0.73)	

(*Continued*)

Table 7.4. (*Continued*)

Developing Country	Dependent Variable	Independent Variables	Chi-Square Value (Prob)	Remarks	Developed Country	Chi-Square Value (Prob)	Remarks
Morocco	PCFW	PCAGDP	10.34 (0.00)	PCAGDP → PCFW	Portugal	8.02 (0.01)	PCAGDP → PCFW
		PCGDP	3.35 (0.18)	–		1.50 (0.47)	
		All	11.13 (0.02)	PCAGDP, PCGDP → PCFW		9.66 (0.05)	PCAGDP, PCGDP → PCFW
South Africa	PCFW	PCAGDP	2.34 (0.30)		Spain	2.82 (0.24)	
		PCGDP	2.51 (0.28)	–		0.75 (0.68)	
		All	3.20 (0.52)	–		3.20 (0.52)	
Sri Lanka	PCFW	PCAGDP	3.78 (0.15)	–	Sweden	6.60 (0.03)	PCAGDP → PCFW
		PCGDP	8.85 (0.01)	PCGDP → PCFW		0.65 (0.72)	
		All	10.57 (0.03)	PCAGDP, PCGDP → PCFW		6.63 (0.15)	
Thailand	PCFW	AO	1.38 (0.50)	–	Switzerland	0.77 (0.96)	
		GDP	0.51 (0.77)	–		0.18 (0.91)	
		All	2.07 (0.72)	–		0.20 (0.99)	
Turkey	PCFW	PCAGDP	0.37 (0.83)	–	United Kingdom	6.05 (0.05)	PCAGDP → PCFW
		PCGDP	3.76 (0.15)	–		1.83 (0.39)	
		All	3.96 (0.41)	–		7.30 (0.12)	
Uruguay	PCFW	PCAGDP	8.60 (0.01)	PCAGDP → PCFW	USA	0.09 (0.95)	
		PCGDP	1.05 (0.58)	–		0.59 (0.74)	
		All	9.15 (0.05)	PCAGDP, PCGDP → PCFW		0.60 (0.96)	

Source: Authors' computations.

Note: '–' → No causal relationship is found.

The probable cause of food waste in developing countries is the poor performance of storage, transport and cooling facilities in different climatic situations. Financial, managerial and technical limitations in harvesting techniques are some main causes for post-harvest loss in developing countries. So here is the involuntary food waste due to lack of storage and poor logistical support (i.e. infrastructure, packaging and marketing system).

Approximately, 700 people are living under undernourishment throughout the world. If we can provide these foods which have been wasted to those people who have no adequate food to live and grow, it can contribute to GDP growth of a country. On the other hand, approximately seven billion tons of food is wasted annually which can also increase the AO of the country as well as of the whole world (Jayadevan, 2022).

Conclusion

With the objective of examining whether PCAGDP and PCGDP maintain long-run and short-run linkages with the food waste out of the cross-border supply chain the following observations are found: in majority of the countries the food waste is associated with PCAGDP and PCGDP of the developing and developed countries, positive for majorly developing and negative for majorly developed countries. In seven out of the twenty developed and eleven out of the twenty developing countries' cases, the three indicators maintain long-run relations which means level of development as measured by PCGDP and PCAGDP have relations with food waste. Food waste is thus an issue of developed countries in general.

Policies framed by government of the countries and world authorities can play an important role on Food Waste reduction. Global food waste policies can be focused on approaches such as legislation in national strategic level, food donation and food waste recycling. Awareness through education and other mediums is also needed. Data collection for food waste is very essential for implementing policies. Promoting social governance and setting up of food banks can contribute global environmental friendliness. Strengthening scientific research projects and case studies on food waste may contribute to the global sustainable development process

References

Alexandratos, N., & Bruinsma, J. (2012). *World agriculture towards 2030/2050: The 2012 revision*. ESA Working paper No. 12-03.

Attenborough, D. (2012). Trying to tackle famine with bags of flour is 'barmy'. *The Guardian*. https://www.theguardian.com/global-development/2013/sep/18/david-attenborough-famine-population

Cervantes-Godoy, D., & Dewbre, J. (2010). Economic importance of agriculture for poverty reduction. Food, Agriculture and Fisheries Working Papers, No. 23.

Chambers, R., & Conway, G. (1992). *Sustainable rural livelihoods: Practical concepts for the 21st century*. Institute of Development Studies.

Chmieliński, P., Pawłowska, A., & Bocian, M. (2023). On-farm or off-farm? Diversification processes in the livelihood strategies of farming families in Poland. *Social Sciences & Humanities Open*, *8*(1), 100575.

Cuéllar A. D. & Webber M. E. (2010). Wasted food, wasted energy: The embedded energy in food waste in the United States. *Environmental Science and Technology*, *44*(16), 6464–6469. doi: 10.1021/es100310d.

Food and Agriculture Organization of the United Nations. (2020, September 29). 15 quick tips for reducing food waste and becoming a Food hero. Food and Agriculture Organization of the United Nations. https://www.fao.org/fao-stories/article/en/c/1309609/

Griffin, M., Sobal, J., & Lyson, T. A. (2009). An analysis of a community food waste stream. *Agriculture and Human Values*, *26*, 67–81.

Gunasena, H. P. M. (2003). Food and poverty: Technologies for poverty alleviation. *South Asia conference on technologies for ooverty reduction*, New Delhi, 10 Ŕ11 October, 2003. https://www.researchgate.net/publication/341867869

Jayadevan, C. M. (2022). Impacts of food wastage on economic growth. *World Food Policy*, *8*(1), 118–125. https://doi.org/10.1002/wfp2.12038

Michael, L., Julie, L., & Jean-Marc, F. (2003). The effects of irrigation on poverty: A framework for analysis. *Journal of Water Policy*, *5*, 413–427.

Oduro, A. D., & Osei-Akoto, I. (2008). Market participation and rural poverty in Ghana in the era of globalization. *African Development Review*, *20*(1), 94–114.

Özerol, G., Bressers, H., & Coenen, F. (2012). Irrigated agriculture and environmental sustainability: An alignment perspective. *Environmental Science and Policy*, *23*, 57–67.

Parfitt, J., Barthel, M., & Macnaughton, S. (2010). Food waste within food supply chains: quantification and potential for change to 2050. *Philosophical Transactions of the Royal Society B: Biological Sciences*, *365*(1554), 3065–3081. https://doi.org/10.1098/rstb.2010.0126

Pothukuchi, K., & Kaufman, J. L. (2000). The food system. *Journal of the American Planning Association*, *66*(2), 113–124.

Rodrigues-Fortes, A., Ferreira, V., Barbosa Simões, E., Baptista, I., Grando, S., & Sequeira, E. (2020). Food systems and food security: The role of small farms and small food businesses in Santiago Island, Cabo Verde. *Agriculture*, *10*(6), 216.

Sheahan, M., & Barrett, C. B. (2017). Review: Food loss and waste in Sub-Saharan Africa. Food Policy, *70*, 1–12.

Singh, A. K. (2013). Income and livelihood issues of farmers: A field study in Uttar Pradesh. *Agricultural Economics Research Review*, *26*, 89–96.

Solis, E. A. (2019). *Food waste reduction in the United States* (Doctoral dissertation).

Tielen, J., & Candel, J. (2014). Reducing food wastage, improving food security? Food & Business Knowledge Platform. https://www.researchgate.net/publication/263747813_Reducing_food_wastage_improving_food_security

Weber, E. P., & Khademian, A. M. (2008). Wicked problems, knowledge challenges, and collaborative capacity builders in network settings. *Public Administration Review*, *68*(2), 334–349.

WTO Committee on Agriculture (2023). *The role of international trade in promoting food security thematic seminar on food security*, FAO of the United Nations. https://www.wto.org/english/news_e/news23_e/agri_31mar23_pres1_e.pdf

Chapter 8

The Anchor Borrowers' Programme and Its Influence on Rice Farming Dynamics in Ebonyi State, Nigeria

Obinna S. Chima, Daniel E. Gberevbie, Moses M. Duruji and Ugochukwu D. Abasilim

Department of Political Science and International Relations, Covenant University, Nigeria

Abstract

This study examines the improvement of rice production in Nigeria through the government's policy of the Anchor Borrowers' Programme (ABP). Rationalized by the production theory, the study utilizes quantitative methods such as surveys. The study investigates the influence of ABP on rice production, with particular emphasis on the experiences of smallholder rice farmers in Ebonyi State, Nigeria. The study employed a cross-sectional survey research approach, and the target population comprised smallholder rice farmers in the state who have benefitted from ABP. The study obtained a representative sample of 400 respondents through stratified random sampling. The study adopts the administration of a structured questionnaire with rice farmers in Ebonyi State, Nigeria, as the primary data collection method. Regression was used to test the hypotheses in SPSS version 23. From the findings, the study reveals the need for the ABP, banks and other lending agencies/programmes to train farmers. It shows that modern farming contributes to rice production in Ebonyi State.

Keywords: Agricultural development; Anchor Borrowers' Programme; Ebonyi state; modern farming techniques; Nigeria; rice production

Sustainable Agricultural Practices: Economic and Environmental Implications, 115–130

doi:10.1108/978-1-83608-336-820241008

Introduction

Rice production is central in Nigeria's agricultural landscape, serving as a primary staple crop and a significant livelihood source for numerous smallholder farmers. Nonetheless, the Nigerian rice sector has long grappled with multifaceted challenges, including suboptimal productivity, outdated farming practices and inadequate access to modern agricultural technologies. In response to these difficulties, the Nigerian government launched the Anchor Borrowers' Programme (ABP) as a pivotal component of a holistic strategy aimed at reinvigorating the agricultural sector and fortifying food security. The ABP strives to provide smallholder farmers with both financial support and technical expertise, enabling them to embrace modern farming methodologies and improve their agricultural yield.

Ensuring food security across various regions of Africa remains a formidable challenge, as the continent's food systems endeavour to provide a significant proportion of its populace with secure, affordable and nutritious diets (Adesina, 2022). Tragically, hunger has become entrenched in the lives of many Africans due to persistent food shortages, rendering the procurement of sustenance a daunting task (Joshi, 2019). Notably, nearly 250 million individuals across the continent endure chronic malnutrition, with an alarming 56 million children under the age of five facing the spectre of starvation (Adesina, 2022). The protracted delay in addressing these imperatives has perpetuated food insecurity, exacerbated by inadequate access to quality seeds, suboptimal fertilizer utilization and ineffective production techniques (Saheed et al., 2018). The confluence of challenges compounded by the COVID-19 pandemic, along with the activities of criminal elements, including bandits and terrorists, has further exacerbated Nigeria's food security predicament, disrupting the food value chain and exacerbating market instability (Oyelami et al., 2023; United States Agency for International Development, 2021).

Regrettably, Nigeria's agricultural sector, envisioned as a linchpin for ensuring food security, has endured protracted neglect (Falayi, 2019), exacerbated by adversities such as the conflict in Ukraine, the COVID-19 pandemic and rampant insecurity in the nation's northeastern region. Despite its pivotal role in sustaining livelihoods for over 70 percent of Nigeria's populace, and its potential to bolster agro-allied industries and amass foreign exchange reserves, the sector has largely been relegated to subsistence-level farming, primarily overseen by elderly rural inhabitants (Balogun et al., 2022). The ramifications of these challenges are starkly evident, with up to 5.1 million individuals in Nigeria at risk of acute food insecurity due to the deleterious activities of criminal syndicates and exacerbated by the COVID-19 pandemic (Alake, 2021; Balogun et al., 2022).

To address the intricate tapestry of challenges confronting Nigeria's agricultural landscape and bolster the nation's pursuit of food security, the ABP emerged under the stewardship of the Central Bank of Nigeria (CBN). This initiative underscores the pivotal role of agricultural and rural development in advancing national development objectives. Against this backdrop, this study explores rice production dynamics within Ebonyi State and its consequential impact on Nigeria's food security paradigm. Ebonyi State, strategically selected due to its prominence in rice processing and marketing, serves as the focal point for evaluating the efficacy of the ABP in augmenting rice production. The study aims to delineate the ABP's influence on rice production, with particular emphasis on the experiences of smallholder rice farmers in Ebonyi State, Nigeria.

Literature Review

We present some reviews on the studies related to the scenario of rice production in the Ebonyi State in Nigeria and the modern farming techniques followed by the farmers in the said state.

Rice Production in Ebonyi State, Nigeria

Nigeria's rice production is a crucial pillar of the country's agricultural landscape and economy. Despite Nigeria being one of Africa's most populous nations, endowed with favourable ecological conditions for rice cultivation, the nation heavily relies on rice imports to meet domestic demand. This reliance on imported rice poses significant challenges to food security and economic stability, underscoring the urgent need to bolster local rice production. In Nigeria, rice has become a strategic food staple, often viewed as a political commodity (Yusuf et al., 2020). Increasing local rice production is paramount for ensuring food security, providing alternative income sources for farmers, bolstering foreign exchange earnings and combating poverty by generating employment opportunities (Ike & Umar, 2011).

However, rice production in Nigeria grapples with a myriad of challenges. Inadequate infrastructure, limited access to credit facilities and pest infestations are significant obstacles to rice farming (Chukwudalu et al., 2021). Moreover, barriers such as educational disparities and restricted access to extension services hinder the adoption of improved rice varieties among smallholder farmers (Oyetunde-Usman et al., 2021). The exacerbation of these challenges by climate change necessitates adopting climate-smart agricultural techniques to mitigate its adverse effects on rice yields (Das & Ansari, 2021). Additionally, fluctuations in global rice prices and government policies on rice importation influence the local rice market, thereby shaping production decisions (Live Rice Index [LRI], 2016). Despite these obstacles, Nigeria boasts immense potential for rice production, with various ecologies suitable for cultivation.

However, the exploitation of these ecologies remains suboptimal, resulting in a significant gap between potential and actual rice production per hectare (Yusuf & Omodara, 2023). Furthermore, rice production is predominantly dominated by smallholder farmers employing traditional farming methods and facing limited access to modern agricultural technologies and inputs (Bukar Adam & Bakshi, 2007). To confront these challenges and bolster agricultural productivity, particularly in rice production, initiatives such as the ABP have been rolled out. Launched by the CBN in 2015, the ABP aims to collaborate with anchor companies engaged in agricultural production and processing to empower rural farmers to ramp up production and supply while curbing importation (Obayelu & Oyinbo, 2020). Given its strategic importance in rice processing and marketing, Ebonyi State was chosen as the study area to evaluate rice production and its contribution to food security since the inception of the ABP.

Modern Farming Techniques and Promotion of Rice Production in Ebonyi State

The Federal Government's ABP is a pivotal initiative in Nigeria's agricultural development and rural prosperity quest. Launched as part of a broader strategy to diversify the Nigerian economy and foster a robust agricultural base, the ABP aims to integrate value chains, bolster food security, reduce imports and enhance productivity (Chioma, 2016). With a specific focus on rice, the ABP seeks to curtail Nigeria's substantial annual import bill for rice and other agricultural commodities, thereby conserving foreign exchange reserves (Central Bank of Nigeria, 2016). In addition to promoting self-sufficiency in food production, the ABP endeavours to drive economic diversification by shifting focus from oil-dependent industries to agriculture, manufacturing and services. The Programme's objectives include fostering an ecosystem that connects out-growers with integrated millers, leveraging modern farming techniques, such as mechanization and precision agriculture, to boost domestic rice production and increasing the operational capacity of integrated rice mills (CBN, 2016).

Furthermore, the ABP seeks to expand access to bank financing for rice production, enhance the capacity of smallholder farmers and prioritize commodities in which Nigeria holds comparative advantages. The ABP operates in alignment with the developmental mandate of the CBN, aiming to stimulate output and mitigate the adverse effects of food importation on the nation's foreign reserves. By providing farm inputs and financial support to smallholder farmers, the programme incentivizes local rice production, fosters collaboration between farmers and agro-processors and addresses the country's negative balance of payment on food items (Edafe et al., 2023; Osabuohien et al., 2020). Rice, the pilot crop for the ABP, holds immense significance in Nigeria's agricultural landscape and economy.

As a staple food for over 50% of the world's population, including a substantial portion of Nigeria's populace, rice occupies a central position in global food security (Akpokodje et al., 2001), In Nigeria, rice cultivation is a major source of employment and income for rural communities, particularly in states like Ebonyi. Despite its historical roots dating back to 1500 BC, rice production in Nigeria has undergone significant transformations. Introducing high-yielding rice varieties and advancements in cultivation techniques and agricultural practices have remarkably increased rice output and productivity (Imolehin & Wada, 2000). However, challenges such as limited access to credit, inadequate infrastructure and climatic uncertainties impede the sector's full potential. Adopting modern farming techniques has become imperative to address these challenges and capitalize on emerging opportunities.

Modern farming techniques encompass diverse practices and technologies designed to optimize agricultural productivity while minimizing environmental impact. These techniques leverage scientific advancements, technological innovations and sustainable farming practices to enhance crop yields, improve resource use efficiency and promote ecological sustainability. Precision agriculture is a critical modern farming technique that harnesses GPS, sensors and drones to optimize crop management practices. By precisely targeting inputs such as fertilizers, pesticides and water-based on real-time data and field variability, precision agriculture enables farmers to maximize yields while minimizing resource wastage (Khosla, 2019). Mechanization, another vital modern farming technique, involves using machinery and equipment for various agricultural operations, including land preparation, planting and harvesting.

Mechanized equipment such as tractors and combine harvesters increases operational efficiency, reduces labour requirements and enhances productivity (Giller et al., 2009). In addition to precision agriculture and mechanization, modern irrigation management techniques are crucial in optimizing water use efficiency and mitigating the adverse effects of water scarcity on crop yields. Technologies such as drip irrigation and precision irrigation enable farmers to deliver water directly to the root zone of crops, minimizing water wastage and maximizing crop productivity (Lamm et al., 2012). Genetic engineering represents another modern farming technique that has revolutionized crop breeding and genetic improvement. Through hybridization and genetic modification, scientists can develop high-yielding crop varieties with desirable traits such as disease resistance, pest tolerance and improved nutritional content (Khush, 2013).

Integrated pest management (IPM) and conservation agriculture are integral to modern farming practices. IPM emphasizes using multiple control tactics, including biological control, cultural practices and judicious use of pesticides, to manage pest populations while minimizing environmental risks (Barzman et al., 2015). Conservation agriculture, on the other hand, promotes minimal soil disturbance, permanent soil cover and diversified crop rotations to improve soil health, conserve water and enhance biodiversity (Reicosky, 2020). Furthermore, agroforestry represents a sustainable farming practice that integrates trees and shrubs into agricultural landscapes to provide multiple benefits, such as soil conservation, carbon sequestration and biodiversity conservation (Pantera et al., 2021). By diversifying farm landscapes and enhancing ecosystem services, agroforestry contributes to sustainable agricultural production and resilience to climate change.

Promoting rice production in Ebonyi State, Nigeria, through initiatives like the ABP and adopting modern farming techniques, holds immense potential for enhancing agricultural productivity, improving food security and fostering rural development. By leveraging precision agriculture, mechanization, genetic engineering and other modern farming practices, smallholder farmers in Ebonyi State can optimize resource use efficiency, increase crop yields and enhance livelihoods. Furthermore, integrating sustainable farming practices such as conservation agriculture and agroforestry can promote environmental sustainability and resilience to climate change, ensuring the long-term viability of rice production in the region.

Based on the foregoing, this study was guided by the following hypotheses:

a. Import substitution does not have a significant contribution to rice production in Ebonyi State and
b. Modern farming techniques contribute little to rice production in Ebonyi State, Nigeria.

Theoretical Framework

Production theory was adopted as the theoretical framework of analysis, aiming to establish the correlation between output and factors of production, mainly focusing on the supply side of product price determination. It elucidates how companies determine the optimal combination of factors to maximize output or minimize production costs. As Edwin (2015) noted, James Mill was among the first economists to analyse production comprehensively, dedicating a chapter to the concept in his seminal work 'Elements of Political Economy'. This theory applies to individual firms and an economy's entire structure or specific economic sectors. The development of production theory has been shaped by the contributions of renowned economists such as Adam Smith, David Ricardo, John Stuart Mill, Alfred Marshall and more contemporary scholars like Paul Samuelson and William Baumol.

These scholars have refined and expanded the theory, enriching our understanding of production processes and their economic implications. Production theory rests on several fundamental assumptions:

a. Rationality: It assumes that producers are rational decision-makers striving to maximize profits within given constraints.
b. Fixed and Variable Inputs: The theory categorizes inputs into fixed (e.g. capital and land) and variable (e.g. labour and raw materials), recognizing their distinct roles in the production process.
c. Law of Diminishing Marginal Returns: This principle posits that as the quantity of one input increases while other inputs are held constant, the marginal output will eventually diminish.
d. Perfect Competition: Many formulations of production theory assume ideal competition, wherein firms are price-takers and face horizontal demand curves for their output.
e. Technological Efficiency: Producers aim to achieve technological efficiency by employing the most practical combination of inputs to produce a given output level.

In the context of this study, production theory provides a relevant framework for understanding the dynamics of rice production in Ebonyi State, Nigeria. The study can analyse factors influencing rice output, input utilization and overall production efficiency by applying production theory. For instance, the law of diminishing marginal returns helps explain how changes in input levels affect rice yields in the state. Moreover, the assumption of rationality allows for examining producers' decision-making processes regarding resource allocation and output optimization. Overall, production theory offers a systematic approach to analysing the production process in agricultural settings. It is well-suited for investigating rice production in Ebonyi State and identifying avenues for enhancing agricultural productivity and efficiency.

Methodology

The study employed a cross-sectional survey research design to evaluate food security and the effectiveness of the ABP in Ebonyi State, Nigeria, with a specific focus on rice smallholder farmers benefiting from the ABP, aiming to address pertinent issues Participant selection utilized multistage sampling techniques, combining two or more probability sampling methods, to form a comprehensive sampling approach (Alvi, 2016; Chidiebere-Mark et al., 2019). The study employed stratified random sampling to identify strata among rice-producing farmers in the local government districts of Ivo, Onicha and Abakaliki in Ebonyi State. The random sampling was used to choose responders from each stratum. The study used a structured questionnaire and in-depth oral interviews to gather information on the population demographics, research questions, food security, improving employment opportunities, poverty reduction and the shift of rural smallholder farmers from subsistence to commercial rice farming.

Ebonyi State, situated in southeastern Nigeria, has become a significant agricultural centre due to its favourable climatic conditions and abundant natural resources (Ofoke, 2018). The state's agricultural sector, particularly rice production, has been influenced by its diverse topography and plentiful water bodies (Eze et al., 2019). Most of the state's population uses subsistence farming, making agriculture the cornerstone of its economy (Onwe et al., 2020). The agricultural landscape of Ebonyi State is characterized by smallholder farmers who rely on both rain-fed and irrigated farming systems for crop cultivation (Okpachu et al., 2017). Government interventions, such as providing agricultural inputs and extension services, have supported rice farmers and fostered agricultural development (Norton & Alwang, 2020). Additionally, efforts to enhance infrastructure, including road networks and irrigation facilities, have contributed to the sector's growth (Ogbonna & Okafor, 2019).

Furthermore, Ebonyi State has benefitted from government-led initiatives to boost rice production and promote agribusiness development (Nwahia, 2020). These interventions, combined with collaboration with development partners, have accelerated agricultural productivity and improved rural livelihoods in the state (Eneh & Nwibo, 2018). Consequently, Ebonyi State has become a key player in Nigeria's quest for food security and agricultural sustainability (Omeje et al., 2021).

Results and Analysis

Table 8.1 presents the information on the demographics of respondents from rice smallholder farmers in Ebonyi State, Nigeria.

Based on the data presented in Table 8.1, the analysis reveals that respondents aged between 18 and 25 constitute 142 individuals (35.5%), those aged between 26 and 35 represent 107 respondents (26.8%), while the age group of 36 and 45 comprises 98 respondents (24.5%). Furthermore, respondents aged 45 and 55 account for 25 respondents (6.3%) and those aged 56 and above constitute 28 respondents (7.0%). Regarding gender distribution, male respondents outnumber female respondents, with 263 respondents (65.8%) and 137 respondents (34.3%), respectively.

Table 8.1. Demographics of Respondents from Rice Smallholder Farmers in Ebonyi State, Nigeria.

Characteristics	Frequency (*N* = 400)	Percentage (%)
Age		
18–25 years	142	33.5
26–35 years	107	26.8
36–45 years	98	24.5
46–55 years	25	6.3
56 and above	28	7
Gender		
Male	263	65.8
Female	137	34.3
Marital status		
Single	242	60.5
Married	112	28
Divorced	39	9.8
Widowed	7	1.8
Educational qualifications		
SSCE	75	18.8
OND	152	38
HND	161	40.3
B.Sc	10	2.5
Postgraduate	2	0.5
Beneficiary of ABP		
Non-beneficiary	64	16
Beneficiary	336	84
Period of benefit from ABP		
0–1 year	52	13
2–3 years	115	28.8
4 years and above	167	41.8
Not applicable	66	16.5

Source: Survey data, 2023.

Regarding marital status, most respondents are married, totalling 242 respondents (60.5%), followed by single respondents, accounting for 112 respondents (28%). Additionally, 39 divorced respondents (9.8%) and 7 widowed respondents (1.8%) were among them.

In terms of educational qualifications, the data indicate that respondents with a Senior Secondary School Certificate comprise 75 respondents (18.8%), while those with an (OND) constitute 152 respondents (38%). Furthermore, respondents with a HND account for 161 respondents (40.3%), whereas those with a Bachelor of Science (BSc) degree represent 10 respondents (2.5%). Finally, respondents with a postgraduate degree comprise 2 respondents (0.5%).

The data indicates that there are 64 non-beneficiaries (16%) and 336 beneficiaries (84%) of the ABP.

Regarding the ABP benefit period, farmers who have benefitted for 0–1 years account for 52 individuals (13%), while those who have benefitted for 2–3 years represent 115 individuals (28.8%). Additionally, farmers who have benefitted for four years comprise 167 individuals (41.8%), while 66 individuals (16.5%) reported that the duration did not apply to them.

Test of Hypotheses

This study examined two hypotheses: Firstly, import substitution has no significant influence on rice production in Ebonyi State. Secondly, a positive relationship exists between modern farming practices and rice production in Ebonyi State, Nigeria.

Linear regression analysis was used to assess the first hypothesis, while correlation analysis was employed to examine the second hypothesis. The findings for the first hypothesis are presented in Panel A of Table 8.2, and those for the second hypothesis are presented in Panel B.

From Panel A of Table 8.2, the influence of import substitution on rice production can be expressed as

$$RP = 0.503 + 0.497 \times IS$$

The findings indicate that for each unit increase in the import substitution program in Ebonyi State, rice production experiences a positive influence of 0.497. Regarding model performance, the R-squared value is 0.649, indicating that the predictor variables explain 64.9% of the variability in the response variable (ABP). The adjusted R-squared value is 0.647. The hypothesis is rejected with an F-statistic of 366.37 at (2,397) degrees of freedom, yielding a P-value of 0.000, less than the significance level of 0.01. The rejection of the null hypothesis at the 1% significance level provides strong evidence that import substitution influences rice production in Ebonyi State.

The Pearson product-moment correlation coefficient was used to test the second hypothesis (H2). The outcome is displayed in Table 8.2, Panel B. The findings reveal a positive relationship between modern farming and rice production in Ebonyi state, Nigeria, with a Pearson product-moment correlation coefficient of 0.631 and a P-value of 0.000, less than 0.01. Consequently, the hypothesis that modern farming does not contribute to rice production in Ebonyi State, Nigeria, is invalidated. It can be concluded that modern farming, involving using portable mechanical tools and the application of fertilizer and farm inputs, has a positive impact on rice production in Ebonyi State, Nigeria.

Table 8.2. Results of Regression and Correlation Analyses for Import Substitution's Influence on Rice Production in Ebonyi State.

a. Panel A: Results of Linear Regression Analysis for Import Substitution's Influence on Rice Production in Ebonyi State, Nigeria

Model		Unstandardized Coefficients		Standardized Coefficients	*T*	Sig.
		β	Std. Error	β		
1	(Constant)	0.503	0.115		4.384	0.000
	Import substitution	0.497	0.040	0.525	12.334	0.000

b. Panel B: Correlation Analysis for Modern Farming's Relationship with Rice Production in Ebonyi State, Nigeria

	Modern Farming	Food Security
Modern	1	0.634**
Farming Pearson Correlation		0.000
Sig. (two-tailed)	400	400
N		
Rice Production Pearson Correlation	0.634**	1
Sig. (two-tailed)	0.000	
N	400	400

Source: Computed by the authors.

Discussion of Findings

The results revealed 347 youths between the ages of 18–45 out of 400 participants were sampled in the survey. The ABP beneficiaries are pretty educated, showing that 280 out of 336 have been to a higher institution with at least an ordinary diploma up to postgraduate studies, as shown in Chapter 4. A comprehensive interview was also conducted and transcribed accordingly. The hypothesis was subjected to regression analysis, and the results showed strong evidence that the import substitution contributed to rice production in Ebonyi State. Below are the highlighted reviews:

Import Substitution and Rice Production in Ebonyi State

The agricultural sector remains crucial for ensuring national food security and alleviating poverty, given that more than 70 per cent of the workforce is employed. Import substitution actively bolsters food security (Egbulonu et al., 2020, p. 295). Hence, this study's primary objective was to investigate import substitution's role in enhancing rice production in Ebonyi State. The findings indicate a notable contribution of import substitution to rice production in Ebonyi State, as evidenced by the *P*-value results. The study underscores the existence of actionable import-substitution policies/strategies to support rice production. Import substitution has served as a mechanism for revitalizing deteriorated food security within the state, promoting local food production by imposing restrictions and preferences on food imports. However, the study also reveals that the government's import substitution strategies still need to be adequately implemented in Ebonyi State. Nonetheless, import substitution has yielded significant benefits for rice production in Ebonyi State.

Similarly, Umeh and Adejo's (2019) study on ABP also uncovers some program benefits, such as employment creation and increased income. Furthermore, Umeh and Adejo (2019) in their research on the assessment of CBN's ABP effects on rice farmers in Kebbi State, Nigeria, found that the program provides credit for farmers and educates them on improved cultural practices through training and extension visits. These improvements in practices have enhanced the entrepreneurship skills and abilities of the farmers. Salisu et al. (2022) also found that ABP shows a positive and statistically significant impact on improvement generation in the agricultural sector. Chioma (2016) also acknowledges that the ABP program has boosted rice production, contributing to a substantial part of the country's importation. It has indeed enhanced agricultural development, most especially the production of rice in the country. The 2016 CBN report also attested that the programme stopped an estimated N3.96 trillion yearly bill on importing rice and other farm produce in Nigeria. The ABP program has provided farm inputs in kind and cash for farm labour to smallholder farmers, which encouraged local production of rice (Belewu et al., 2023), to enhance the capacity utilization of integrated mills, stabilize inputs supply to agro-processors and address the country's negative balance of payment on Food. Smallholder farmers deliver their harvest to the agro-processor (the anchor), who then credits the farmer's account with the monetary equivalent.

Modern Farming Technique and Rice Production in Ebonyi State

Modern farming could enhance farmers' access to funds to facilitate farming activities, improving rice production. This study, therefore, investigated the extent to which modern farming contributes to financial inclusion in Ebonyi state, Nigeria. This research reveals that modern farming contributes to rice production in Ebonyi State, Nigeria ($r = 0.634$ and *P*-value). The result shows that many smallholder farmers in Ebonyi State have expanded their production capacity with the help of modern farming tools and farm inputs; the State of rice production in Ebonyi State today is better than before due to innovative financing. Umeh and Adejo's (2019) study also shows that funding innovation contributes to rice production and improves the living standard of rice farmers, which corroborates the study's results.

Conclusion

The ABP has played a crucial role in fostering the development of a robust ecosystem that promotes value chain financing and provides market assurance to smallholder farmers in Ebonyi State. Through this initiative, the CBN has created millions of direct and indirect jobs, thereby contributing to the state's economic growth and poverty reduction. However, despite its commendable achievements, the ABP faces significant challenges that hinder its effectiveness in achieving the food security objectives it aims to address.

Evidence from the literature and interviews revealed several obstacles, including underdeveloped agricultural infrastructure, poor irrigation facilities, ineffective extension services and the absence of an efficient commodity exchange and warehouse receipt financing system. Additionally, the reluctance of some beneficiaries to repay loans poses a significant challenge to the program's sustainability.

In examining the role of import substitution strategies and modern farming techniques in enhancing rice production in Ebonyi State, contrasting findings emerged. While recognized as a vital strategy for national food security and economic development, import substitution was found to have a limited impact on rice production in the state due to inadequate implementation of policies and strategies. Conversely, modern farming techniques have proven instrumental in enhancing rice production in Ebonyi State. Through innovative financing mechanisms and access to modern farming tools and inputs, smallholder farmers have significantly expanded their production capacity, leading to tangible improvements in rice production levels. This underscores the importance of continued investment in modern agricultural practices and financial inclusion initiatives to boost rice production further and improve farmers' livelihoods.

The ABP has emerged as a beacon of hope, with its positive impact on rice production and farmer livelihoods evident across different states in Nigeria. By providing credit, training and support to smallholder farmers, the ABP has boosted rice production and contributed to poverty alleviation and economic growth. Sustained support for programs like the ABP and strategic interventions to address existing challenges hold the key to unlocking the full potential of rice production in Ebonyi State and beyond. Based on the findings and insights gathered from the study on import substitution, rice production and the role of modern farming techniques in Ebonyi State, Nigeria, the following actionable recommendations are proposed, along with explicit strategies for their achievement:

a. Strengthen Implementation of Import Substitution Policies: The Ebonyi State Government, in collaboration with relevant federal agencies and agricultural experts, should develop a comprehensive import substitution policy framework. Dedicated task forces or committees, comprising representatives from government agencies, farmer associations and civil society organizations, should be established to oversee the implementation of import substitution initiatives. Adequate budgetary allocations should be made to support the enforcement and monitoring of import substitution strategies, with regular progress reports provided to stakeholders.

b. Enhance Access to Agricultural Credit and Training: The CBN, in partnership with microfinance institutions, rural banks and agricultural cooperatives, should expand the reach of agricultural credit programs such as the ABP. Government agencies and non-governmental organizations (NGOs) should organize capacity-building workshops and training sessions on modern farming techniques, financial literacy and business management for smallholder farmers. Utilizing digital platforms and mobile technology, agricultural extension services should be extended to remote or underserved areas, focusing on providing tailored training and support to farmers.
c. Improve Agricultural Infrastructure and Extension Services: The Ebonyi State Government, with support from federal agencies and private-sector partners, should invest in upgrading irrigation infrastructure, farm mechanization equipment and post-harvest handling facilities. Additional extension workers should be recruited, trained and deployed to rural communities to provide on-the-ground technical assistance and knowledge transfer on sustainable agriculture practices. Collaborative research and development projects should be initiated with research institutions and agricultural universities to address specific challenges related to soil fertility, water management and pest control.
d. Foster Market Linkages and Value Chain Integration: Public–private partnerships should be established to develop market information systems and electronic trading platforms, facilitating transparent price discovery and linking farmers with buyers and suppliers. Farmer producer organizations, cooperatives and agribusiness clusters should be actively promoted and supported to strengthen market access and value chain coordination. Investment incentives and regulatory support should be provided to attract agro-processing industries, food processors and agribusiness investors to Ebonyi State.
e. Enhance Public Awareness and Stakeholder Engagement: The Ebonyi State Government, in collaboration with NGOs and community leaders, should conduct targeted communication campaigns through various channels, including radio broadcasts, community meetings and social media outreach. Stakeholder consultations and participatory workshops should be organized to engage farmers, policymakers, civil society organizations and private sector actors in collaborative decision-making and action planning. Knowledge-sharing platforms, such as seminars, conferences and online forums, should be established to facilitate peer-to-peer learning and the exchange of best practices among agricultural stakeholders.

References

Adesina, A. (2022). It is time to nourish Africa once and for all. Retrieved from https://www.afdb.org/en/news-and-events/its-time-nourish-africa-once-and-all-50638

Akpokodje, G., Lançon, F., & Erenstein, O. (2001, November). Nigeria's rice economy: State of the art, In NISER/WARDA Nigerian Rice Economy Stakeholders Workshop, Ibadan (pp. 8–9).

Alake, T. O. (2021). Food security and hunger in Nigeria: Issues, challenges, and policy implications. In M. Adenrele & O. I. Odeyemi (Eds.), *Food security and sustainable development in Nigeria* (pp. 231–248). IGI Global.

Alvi, M. (2016). A manual for selecting sampling techniques in research. *Munich personal RePEc archive*. University of Karachi, Iqra University. https://mpra.ub.uni-muenchen.de/70218/

Balogun, A. L., Adebisi, N., Abubakar, I. R., Dano, U. L., & Tella, A. (2022). Digitalization for transformative urbanization, climate change adaptation, and sustainable farming in Africa: Trend, opportunities, and challenges. *Journal of Integrative Environmental Sciences*, *19*(1), 17–37.

Barzman, M., Bàrberi, P., Birch, A. N. E., Boonekamp, P., Dachbrodt-Saaydeh, S., Graf, B., … & Sattin, M. (2015). Eight principles of integrated pest management. *Agronomy for sustainable development*, *35*, 1199–1215.

Belewu, K. Y., Ajao, A. O., & Ezekiel, A. A. (2023). Rice farmers' technical efficiency and level of poverty: Evidence from the Anchor Borrower Program (ABP). *International Journal of Agricultural Science, Research and Technology in Extension and Education Systems* (IJASRT in EESs), *13*(2), 119–130.

Bukar Adam, S., & Bakshi, J. S. (2007). Potential of improved rice parboiling technology in Nigeria. *Agricultural Engineering International: CIGR Journal*, *9*, 1–12.

Central Bank of Nigeria (2016, December). *Anchor borrowers' programme (ABP) guidelines*. Development Finance Department.

Chidiebere-Mark, N., Ohajianya, D., Obasi, P., & Onyeagocha, S. (2019). Profitability of rice production in different production systems in Ebonyi State, Nigeria. *Open Agriculture*, *4*, 237–246.

Chioma, A. (2016). *Appraising Central Bank of Nigeria Anchor Borrowers' Programme*. Retrieved from http://Reporters.com/2016/07/903611

Chukwudalu, A. C., Nnoli, L., & Orjiakor, N. (2021). Socio-economic and political policies, and the challenges of Governance under Governor Theodore A. Orji of Abia State, Nigeria, 2007–2015. Nigerian Journal of African Studies *(NJAS)*, *2*(2), 1–23.

Das, U., & Ansari, M. A. (2021). The nexus of climate change, sustainable agriculture and farm livelihood: Contextualizing climate smart agriculture. *Climate Research*, *84*, 23–40.

Edafe, O. D., Osabuohien, E., Matthew, O., Olurinola, I., Edafe, J., & Osabohien, R. (2023). Large-scale Agricultural Land Investments and Food Security in Nigeria. *Heliyon*, *9*(9), 1–14.

Edwin, C. (2015). A history of the theories of production and distribution in English political economy. *The Economic Journal*, *3*(11), 1.

Egbulonu, K. G., Duru, E. E., & Echeta, D.O. (2020). Agriculture and food security in Nigeria: The natural resource-course nexus. *Ikoro Journal of the Institute of African Studies*, *10*(1 & 2), 209–304.

Eneh, F. C., & Nwibo, S. U. (2018). Government agricultural policies and rural development in Nigeria: A case of Ebonyi State. *Nigerian Journal of Agriculture, Food and Environment*, *14*(4), 15–20.

Eze, N. M., Onyishi, G. C., & Ugwumba, O. A. (2019). Factors influencing women's involvement in rice processing in Ebonyi State, Nigeria. *African Journal of Agriculture, Food, Nutrition and Development*, *19*(2), 14648–14661.

Falayi, (2019). *Why Nigeria has restricted food imports*. Retrieved from https://www.bbc.com/news/world-africa-49367968,

Giller, K. E., Witter, E., Corbeels, M., & Tittonell, P. (2009). Conservation agriculture and smallholder farming in Africa: The heretics' view. *Field Crops Research*, 114(1), 23–34.

Ike, P. C., & Umar, H. S. (2011). Evaluation of small-scale rice production in Kano state, Nigeria. *Journal of Agricultural Extension and Rural Development*, *3*(10), 181–186.

Imolehin, E. D. (1991). *Rice improvement and production in Nigeria*. The paper was presented at the WARDA upland breeding task force workshop, Bouaké, Côte d'Ivoire.

Joshi, S. (2019). *Food is more than an Essential Need*. Retrieved from https://nutritionmeetsfoodscience.com/2019/01/03/food-more-than-an-essential-need

Khosla, R. (2019). Growth & productivity of food processing industries in India. *Indian Journal of Industrial Relations*, *55*(1), 54–66.

Khush, G. S. (2013). Strategies for increasing the yield potential of cereals: Case of rice as an example. *Plant Breeding*, *132*(5), 433–436.

Lamm, F. R., Stone, L. R., & Ahuja, L. R. (2012). Deficit irrigation scheduling of maize and potatoes using soil moisture measurements. *Transactions of the ASABE*, *55*(4), 1373–1386.

Live Rice Index [LRI]. (2016). Global rice prices & latest news. https://livericeindex.com/

Norton, G. W., & Alwang, J. (2020). Changes in agricultural extension and implications for farmer adoption of new practices. *Applied Economic Perspectives and Policy*, *42*(1), 8–20.

Nwahia, O. C. (2020). Analysis of the cost and economic returns in rice production in Ebonyi State, Nigeria. *Indonesian Journal of Agricultural Research*, *3*(3), 205–214.

Obayelu, A. E., & Oyinbo, O. (2020). Analysis of the effect of anchor borrowers' programme on rice production in Nigeria. *International Journal of Agricultural Economics and Rural Development*, *3*(1), 112–124.

Ofoke, M. O. (2018). Assessment of rice production and utilization in Ebonyi State, Nigeria. *Journal of Agricultural Extension and Rural Development*, *10*(2), 14–28.

Ogbonna, O. S., & Okafor, F. C. (2019). Analysis of agricultural infrastructure development in Ebonyi State, Nigeria. *Journal of Agricultural Science and Research*, *2*(1), 18–27.

Okpachu, G., Akpa, A. D., & Nnanna, U. (2017). Evaluation of rice farming systems in Ebonyi State, Nigeria. *Journal of Agriculture and Rural Development in the Tropics and Subtropics*, *118*(1), 79–92.

Omeje, O., Eze, A. N., & Eze, C. C. (2021). Impact of government interventions on rice production in Ebonyi State, Nigeria. *Journal of Agricultural Economics and Development*, *5*(2), 67–78.

Onwe, I. J., Eze, J. I., & Ewa, E. E. (2020). Sustainable agricultural development in Ebonyi State, Nigeria: Challenges and prospects. *International Journal of Agricultural Extension*, *8*(1), 34–47.

Osabuohien, E. S., Olokoyo, F. O., Efobi, U. R., Karakara, A. A., & Beecroft, I. (2020). Large-scale land investments and household livelihood in Nigeria: Empirical insights from quantitative analysis. In *The Palgrave Handbook of Agricultural and Rural Development in Africa* (pp. 133–152). Springer.

Oyelami, L. O., Edewor, S. E., Folorunso, J. O., & Abasilim, U. D. (2023). Climate change, institutional quality, and food security: Sub-Saharan African experiences. *Scientific African*, e01727.

Oyetunde-Usman, Z., Olagunju, K. O., & Ogunpaimo, O. R. (2021). Determinants of adoption of multiple sustainable agricultural practices among smallholder farmers in Nigeria. *International Soil and Water Conservation Research*, *9*(2), 241–248.

Pantera, A., Mosquera-Losada, M. R., Herzog, F., & Den Herder, M. (2021). Agroforestry and the environment. *Agroforestry Systems*, *95*(5), 767–774.

Reicosky, D. (2020). Conservation agriculture systems: Soil health and landscape management. In *Advances in conservation agriculture* (pp. 87–154). Burleigh Dodds Science Publishing.

Saheed, Z. S., Alexander, A. A., Isa, A. A., & Adeneye, O. A. (2018). Anchor borrower proramme on agricultural commodity price and employment generation in Kebbi state, Niger. *European Scientific Journal*, *14*(13), 240–255.

Salisu, I., Adebayo, C. O., Jirgi, A. J., & Ojo, A. O. (2022). Effects of Anchor Borrowers Programme (ABP) credit on the productivity of beneficiary rice farmers in Kebbi State, Nigeria. *FUDMA Journal of Agriculture and Agricultural Technology*, *8*(1), 329–338.

Umeh, J. C., & Adejo, M. A. (2019). *Assessment of central bank of Nigeria's anchor borrowers' programme effects on rice farmers in Kebbi State*, Nigeria.

United States Agency for International Development (2021). *USAID/Nigeria Covid-19 food security challenge*. Retrieved from USAID/Nigeria COVID-19 Food Security Challenge U.S. Agency for International Development.

Yusuf, H., & Omodara, O. J. (2023). Climate change, rice production, and food security in Nigeria: A review. *Journal of Agricultural Science and Technology*, *25*(1), 1–14.

Yusuf, W. A., Yusuf, S. A., Adesope, A. A. A., & Adebayo, O. Z. (2020). Determinants of rice import demand in Nigeria. *Journal of Applied Sciences and Environmental Management*, *24*(5), 923–931.

Chapter 9

Zero Budget Farming and Export Competitiveness of Agricultural Sector in LDCs: An Introspection with Reference to India

Debashis Mazumdar[a] *and Mainak Bhattacharjee*[b]

[a]*Department of Economics, The Heritage College, Kolkata, India*
[b]*Department of Economics, Loreto College, Kolkata, India*

Abstract

Introduction of organic farming and reduced dependence on inorganic family has positive implications for the economic viability of the small farming units in India's agricultural sector. The privatization of access to seeds, fertilizers and other inputs, as an outcome of economic liberalization, has deeply dented the affordability of agriculture, especially at the level of small and micro farms, thereby undermining the export potential of the agricultural sector as well. Zero budget farming can be potentially effective in weathering out the affordability crisis and restoring economic viability of farming at large. This chapter is a moderate attempt to explore how a shift towards zero-budget farming from inorganic farming can improve the access of agricultural goods, as predominant component of their export basket, to the markets in developed countries (DCs), given the rising stringency of sanitary and phytosanitary and environmental standards, as well, coming disproportionately against the exports from developing and less-developed countries (LDCs), including India. With reference to Balassa's the measure of revealed comparative advantage, the chapter points to the dwindling competitive advantage of India's agriculture in global market during the last decade. Given the stringent environment standard facing, disproportionately, the exports from LDCs, zero-budget

Sustainable Agricultural Practices: Economic and Environmental Implications, 131–143

doi:10.1108/978-1-83608-336-820241009

farming can improve the access of agricultural goods, as predominant component of their export basket, to the markets in DCs. The benefit from the creation of public goods by means of organic farming practices can be utilized to deepen the fiscal support to small and marginal farms.

Keywords: Zero-budget natural farming (ZBNF); economic liberalization; sanitary & phitosanitary standard (SPS); developing and less developed countries (LDC); export competitiveness; environmental standards; organic farming

Introduction

Agricultural growth strategy that has been followed in India over the past few decades clearly reveals that during 1960s, the 'new technology strategy' in terms of a 'seed-fertilizer-water' package or a 'Green Revolution Model' emphasized on greater application of chemical fertilizers, high yielding variety seeds, chemical pesticides along with proper irrigational facilities to raise the production of food grains. This approach no doubt succeeded in increasing food grain production but it did not address the issue of sustainability. Over time, indiscriminate use of chemical fertilizers, underground water for irrigation and pest control measures have led to the decline in soil health and had an adverse impact on productivity, and this trend was combined with growing pollution, desertification and a loss of agro-diversity causing our agricultural activities unsustainable. Increasing costs of inputs and increasing risks, particularly for the small farmers, in undertaking such input-intensive cultivation process, made agricultural practices less remunerative and led to growing incidences of farmers' indebtedness. In 2013, about 52 per cent of the farm households in India were in debt (NSSO, 2013). There is no doubt that privatization of access to seeds, fertilizers and other inputs, as an outcome of economic liberalization, has deeply dented the affordability of agriculture, especially at the level of small and micro farms, thereby undermining the export potential of the agricultural sector as well. Following the 'Green Revolution Model' though it was possible to raise agricultural output in India, this performance was not translated into her export performance in agricultural trade. In 1989, though India's share in world agricultural output was about 10 per cent, her share in world trade in agricultural commodities was only around 0.6 per cent (Nayyar & Sen, 1994). Although India's share in world agri-export has shown an increasing trend during 2010–2022, and though India is among the top 10 agri-exporting countries, her share in world agri-export was only 2.3 per cent in 2022 (WTO, 2023).

The zero-budget natural farming (ZBNF) model was an outcome of the growing concern for sustainable agriculture. It brings down market-based dependence of the farmers, viz. in terms of purchase of inputs which are manufactured in origin as it encourages the use of locally available natural fertilizers and other farming materials. Introduction of ZBNF and reduced dependence on inorganic inputs has positive implications for the economic viability of the small farming units in India's agricultural sector. The ZBNF can be potentially effective in weathering out the affordability crisis and restore economic viability of farming at large in India. This apart, given the stringent environment standard [sanitary and phytosanitary (SPS) measures] facing, disproportionately, the exports from less developed countries (LDCs), ZBNF can improve the access of agricultural goods, as predominant component of their export basket, to the markets in developed countries (DCs). In India, the government of Andhra Pradesh is supposed to be the forerunner among all the states in implementing the ZBNF as it initiated the climate-resilient ZBNF programme in 2016. In fact, this programme was built upon Andhra Pradesh's past experience of Non-Pesticide Management and Community Managed Sustainable Agriculture (CMSA) programmes which remained operative during 2005–2013 (Khurana & Kumar, 2020). Several other states like Himachal Pradesh, Gujarat, Haryana, Karnataka and Kerala have also initiated ZBNF. Several studies have indicated that average cost of cultivation under ZBNF has been lower than non-ZBNF units, and the net revenue of the farmers was found to be higher under ZBNF than those under non-ZBNF units.

However, the introduction of SPS standards on food quality, safety and health standard by the DCs, in the recent times, has proven to be a strong impediment before developing and LDCs, including India, in the way to sustained growth in agricultural export. This is well evident from the recent incidents of export rejection, particularly by countries in European Union, which has been the predominant destination of India's food and agricultural exports. Keeping this view in mind this chapter seeks to delve into the extent the shift to organic and natural farming in zero-budget model from inorganic farming as predominant form of agricultural practice. This would be offering an economically viable way out of the ongoing doldrums in export performance of India's agricultural sector.

Literature Review

Nayyar and Sen's (1994) accounts for the role of traditional exports originate in the agricultural sector, both plantation crops such as tea or coffee and cash crops such as tobacco or spices, that have always been an in-generation foreign exchange earnings for the economy. They have also added that for these commodities, the open trade regime has continued from the colonial era. However, it has been noted herein, that in case of agriculture, foreign trade flows have been perceived as a residual whether it is to do with exportables or importables. Now, it has been further added that ex-post, trade flows always measure the difference between domestic production and domestic consumption after we allow for net changes in inventories. It needs to be stressed that, in India, agricultural trade flows have been a residual in an ex ante sense. This has been implemented either through quantitative restrictions or through canalization or, over time, through some combination of both. The rationale underlying this trade policy regime for agriculture, which regulated the volume of both exports and imports, has been a concern about domestic prices. A large proportion of India's agricultural exports, whether consumer goods or intermediate goods, are exportables which enter domestic consumption or use so that the volume of exports can exercise a strong influence on domestic prices. The current policy regime featuring trade liberalization indicates a case implicit taxation producers of agricultural commodities (and the effective taxation also varies substantially across crops) in contrast with consumers of these commodities enjoying subsidies. The situation is exactly the opposite in much of the industrialized world, particularly in the European Community and Japan, where the policy regime subsidizes producers of agricultural commodities directly from the exchequer or indirectly through implicit taxation of consumers. Hence, implicit subsidies to exports and employment are desirable in a labour surplus, foreign exchange constrained economy, if there is to stand a 'second-best or third- best' efficiency argument for not removing trade controls on agriculture indiscriminately.

Agarwal (1988): Agricultural products and raw materials formed a larger share of India's exports than of world exports during 1965–1980, and world exports of these two categories of goods had been growing slower than the total world exports during that period. However, the declining share of India's exports in the case of agriculture was particularly due to the negative market growth effect consequent upon slow growth of EEC market for agricultural goods during that period. The slower growth of the prices of India's agri-export in relation to the world prices was also the reason behind poor competitiveness of India's ari-exports during that period.

Gill and Brar (1996) contend that India, to be internationally competitive in wheat and rice, will have to increase farm productivity, introduce efficiency in market handling and lower the tax rates in marketing of these crops. With growing consumers' quality consciousness in advanced countries and signing of SPS agreement our crops have to be of high quality with minimum toxic residues.

Vyas (1999) reviews the performance of Indian agriculture in the context of the development in agricultural trade especially agricultural exports internationally and outlines a strategy for augmenting India's agricultural exports. It emphasizes that Indian exporters must be fully aware of the standards accepted by the international importers and should have the mechanism to introduce these standards right at the primary level. (Own comment: Present emphasis on ZBNF seems to be policy in the right direction.)

Gulati (2002) claims that Indian agriculture is gradually being integrated into world agriculture. But the volatility of world prices, especially in its downswing, worries the Indian policymakers as it results in gushing imports and undermines the production base of millions of small and marginal farmers. It is more disturbing when exporting countries subsidize their exports or are able to compensate their farmers for the fall in world prices. The study undertaken here shows that rice and wheat, the biggest crops of Indian agriculture, have been largely competitive on an importable basis, with their protection coefficients being below unity in most of the years during 1990s. However, performance in agricultural front depended largely upon the fluctuations of agricultural prices in the global market.

Lee (2005) highlights the key factors influencing the adoption of sustainable agricultural (SA) technologies and natural resource management (NRM) practices in developing countries. The article then discusses the investments and policies needed to improve the dissemination and adoption of SA and NRM technologies, practices and systems in developing countries. Policies to improve the information and knowledge base are likely to lead to particularly important impacts. Increased investments in demand-led, farmer-centred adaptive research on SA and NRM technologies will help in improving the understanding of the underlying scientific base of integrated nutrient management, integrated pest management and other SA and NRM practices and systems.

Viswanathan et al. (2012) emphasize upon the need for evolving new perspectives and policies towards sustainable and non-disruptive transformation of smallholder agriculture in Asia in view of the growing challenge of climate change and food security. Preliminary evidence suggests that there are obstacles to export-led agricultural growth in many developing countries because of (a) rising production costs due to increasing wages and input costs; (b) technical barriers to trade and new barriers linked to intellectual property rights, animal welfare, and sanitary and phyto-sanitary measures and (c) inertia and inflexibility of subsistence-oriented production systems that are typical of many farming communities. Though countries such as Vietnam and India have been somewhat successful in accessing international markets under liberalized policy regimes, a major share of their exports comprise commercial agri-crops such as coffee, rubber, tea and pepper, the prices of which are highly volatile.

Cheong et al. (2013) show that, for the typical developing country, growth in agricultural productivity may be a potent means for reducing informal employment, relieving urban congestion, and decreasing income inequality. These effects appear to be strongest in developing countries with larger initial shares of the labour force in agriculture and informal employment. In particular, higher agricultural productivity can increase the quality of work and the wages of agricultural workers, reabsorb workers from the informal sector back into the formal agricultural sector, and leave manufacturing employment relatively unaffected. These findings are particularly important for the least DCs and developing countries where informal employment is rampant due to a premature exit from agriculture, and more productive work is unavailable or limited in the manufacturing and services sectors. As for the sources of agricultural productivity improvement, the literature indicates that technical innovations and modern inputs, when properly adapted to developing-country settings, can enhance labour and increase demand for agricultural workers rather than displace them. The effects of agricultural innovation, however, are conditioned by policies that affect the domestic and external markets for agricultural inputs and outputs.

Bargout (2014) shows in Indian agriculture, the smallholders, facing a common rural reality of disempowerment and limited disposable household capital, the agro-ecological results of climate volatility will have catastrophic costs. In a single growing season, even moderate climate abnormalities in temperature regime or moisture regime have shown to disrupt farm agro-ecology and diminish harvests over time. The ecological farming offers the most practical and holistic traits of resilience, particularly in the areas of soil, water and biodiversity. When acknowledging ecological agriculture as an empowering adaptation strategy for smallholders, the evident sustainability of this approach is also apparent alongside key structural dynamics limiting its adoption.

Tripathi et al. (2018) observe that the farmers use four treatments to enhance productivity at low cost: Bijamrita (seed treatments using local cow dung and urine); Jiwamrita (soil inoculants made of cow dung and urine); Mulching (ensure favourable microclimate soil) and Waaphasa (soil aeration). These treatments are chemical-free agriculture at lowest/zero-cost and enhance the incomes of farmers. Besides, the ZBNF utilizes the existing institutional platform of women's self-help groups, for scaling, sustaining and deepening the program. A total of 1,614 crop-cutting experiments (CCEs) conducted by RySS in ZBNF and non-ZBNF crops in 2017 in Kharif crops have shown encouraging results: about 88% of the CCEs have shown increase in net incomes due to increases in yield and decrease in cost of cultivation; increases in yield were observed across all major food crops and cash crops.

Agarwal et al. (2018) present the climate-smart village (CSV) approach as a means of performing agricultural research for development that robustly tests technological and institutional options for dealing with climatic variability and climate change in agriculture using participatory methods. This chapter has indicated that India and Nepal, as partners of the program on Climate Change, Agriculture and Food Security (CCAFS), had adopted the CSV approach. It aims to scale up and scale out the appropriate options and draw out lessons for policy makers from local to global levels. The approach incorporates the evaluation of climate-smart technologies, practices, services and processes relevant to local climatic risk management and identifies opportunities for maximizing adaptation gains from synergies across different interventions and recognizing potential maladaptation and trade-offs.

Laishram et al. (2022) account natural farming as a special form of agriculture that does not require any financial expenditure to purchase the essential inputs such as seeds, fertilizers and plant protection chemicals from the market. Natural farming, though in its preliminary stages, is showing increased positive results and is being adopted by farmers in good faith. Under natural farming system, two or three crops are cultivated on the same farmland. The study further reveals that farmers witnessed a drop in per hectare cost of production and profitable yield for their crops as well. The crop equivalent yield under natural farming was highest in all the crop combinations as compared to conventional farming and ranged from 3.08 to 5.10% in Kharif season and 2.83 to 7.98% in all the crop combinations in Rabi season. In Kharif season, the percentage reduction in cost of cultivation under NF over the CF system ranged from 12.56 to 30.73, while in Rabi season, it ranged from 6.86 to 12.34. The gross returns under NF systems were highest in all the crop combinations as compared to CF systems.

Duddigan et al. (2022) show that there is no short-term yield penalty when adopting ZBNF in small scale farming systems compared to conventional and organic alternatives, based on 20 field plot experiments established across six districts in Andhra Pradesh (SE India), managed by locally based farmer researchers. These findings could be of particular interest to short-term lease tenant farmers. Initial assessment suggests, however, that performance of ZBNF may be agro-climatic context specific. Further work is needed to evaluate the long-term effects of ZBNF adoption. The southern (drier) districts of Andhra Pradesh (Anantapur, Kadapa, Nellore, Prakasam) had the highest yield in the ZBNF compared to the conventional and organic treatments. It is also found that improved water holding capacity could be an important yield-promoting mechanism responsible for improved performance of ZBNF systems in more arid regions. ZBNF is the only system featured in this research that uses mulching, which can regulate soil temperature and moisture to improve crop yield.

Sarah et al. (2023) provide evidence from replicated field trials (carried out in six districts of South India) to assess the performance of ZBNF compared to conventional and organic alternatives, and mechanistically account for the benefits ZBNF can provide. The results of this study, based on a restricted maximum likelihood mixed effects model, suggest that adoption of ZBNF practices provides a significant yield advantage over organic and conventional alternatives. The ZBNF treatment resulted in a significantly ($p < 0.05$) higher yield compared to organic or conventional treatments overall.

Sharma et al. (2023) posit a conflict between the incremental innovation (green revolution) and radical innovation (natural farming). In what constitutes Lewis Trap, the study indicates that 53% of the world population is composed of farmers with micro farms. Data from 1961 to 2013 showed that the lower the land holding, the higher the land productivity. In the 100% natural farming scenario, productivity is slightly lower compared to 100% industrial system scenario. The study further confirms that India has a strong comparative advantage in producing food in symbiosis with nature, market values, investments and entrepreneurship and start-ups upstream to provide a diversity of quality products, as well as services such as carbon farming, multi-utility farm residue-based products and radical innovations in renewable energy use. However, equity and equality for resources sharing and their management at community level will decide the rate of success in natural farming.

A Simple Theoretical Model on Zero-Budget-Natural Farming in a Small and Less Developed Country

This section demonstrates a simple theoretical model developed in partial equilibrium framework to highlight how a structural transformation in agriculture led by the adoption of ZBNF practices can be effective towards reducing environmental cost of inorganic farming as predominant form agriculture at commercial level having adverse welfare implications, while restoring the welfare gain from barrier-free foreign in agriculture goods in face of stringent SPS restrictions. The abstract of agricultural sector upheld in the model is cast to consist of only farming activities, excluding livestock generation, the output which finds its destination in a perfectly competitive market. This particular construct of agricultural sector fits well into the context of LDCs with large population, with a substantial portion of it being dependent on agriculture for livelihood and the farming ecosystem featuring a predominance of small and micro farms. In this model, it has been posted in the first place that the opening up of domestic agriculture to foreign can bode in welfare gains. Now, while such welfare gain gets discounted by the environmental cost of inorganic farming as the dominant form of commercial agriculture, the model identifies the promotion of natural farming at commercial level and is going to be a significant game-changer. Going further, the model shows how the proliferation of organic farming would effectively counter the barrier for agricultural export arising out of the SPS sanctions.

The dispensation noted above has been illustrated in the following diagram:

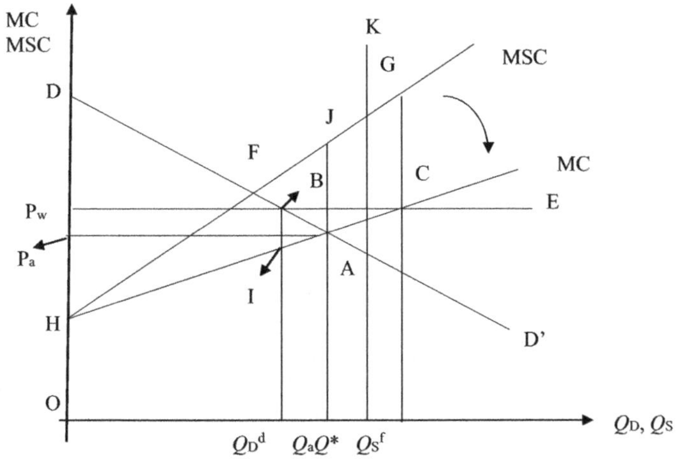

Fig. 9.1. Equilibrium of the Agricultural Sector. *Source*: Drawn by the authors.

In Fig. 9.1, it is being seen that under autarky the equilibrium output of agriculture is Q_a, with the equilibrium price being P_a. Accordingly, the total surplus (measuring efficiency and welfare) arises under autarky is equivalent to the area (ADH). However, given the predominance of inorganic farming, which is highly to the use of chemical fertilizers and pesticides, the social cost of agriculture (measured by the area OHJQ_a) exceeds the private cost (measured by the area OHAQ_a), to an extent such inorganic farming results in damage to environment and consequently, impairs efficiency and welfare, by involving deadweight loss, measured by the area (FAJ). Now post opening up of agricultural sector to foreign trade, domestic price of agricultural good gets aligned with the world market P_w, which is exogenously given to a small and LDC, per se. Now, in the presence of foreign demand, the domestic agricultural sector finds its new demand schedule, as indicated by DBE. According the new equilibrium output of agriculture comes at Q_S^f, of which Q_D^d units go into catering domestic demand at world price (or free trade price) and the rest goes to export. Now, in this free-trade equilibrium, the total surplus is measured by the area (HDBC) of which the part indicate by the area (ABC) represents the welfare gain from free-trade, which however is to be weight against the more environment cost and thereof, a larger amount dead-weight loss, by the area (AFGC). It is at this point the context of natural farming finds relevance. The adoption of natural farming as an alternative to inorganic farming will help reverse the environmental damage resulting from the latter and thereby bring down social cost of agriculture towards its private counterpart, in a way the loss in efficiency and welfare, as already explained, gets replenished. The other end to the benefit of replacing inorganic farming by natural farming is to be traced in the context SPS restriction facing agricultural exports, as a predominant form of non-tariff barrier. Due to such restrictions, a small and LDC will find its exports being constrained at level below its capacity. In terms of the diagrammatic disposition, the equilibrium export volume will fall below the level of $Q_D{}^dQ_S{}^f$ units (say, at $Q_D{}^dQ^*$), and this in turn will dampen the welfare gain from unrestricted trade, as already indicated. Thus, to this end, natural farming can be dubbed as a double engine vehicle, which on the one hand drives the domestic agricultural sector towards efficiency by reducing the environmental cost of production and on the other hand restores its export competitiveness by dint of ensuring easier compliancy with SPS norms and thereby removing the impediment caused otherwise. Hence, to put it in terms of diagram, a switch over from inorganic farming to natural farming will realign the social cost of agriculture with private cost, so as eliminate the inefficiency caused by the dead weight and at the same the reinstate the welfare gain from barrier-free export.

Dynamics of Comparative Advantage and Export Performance of Agricultural Sector in India Post-Economic Liberalization

In this section we will explore the dynamics of export performance of India' s agricultural sector in the period following economic liberalization in comparison with the scenario of the world as a whole. As evident from Panel 1(a) and Panel 1(b) of Fig. 9.2 India witnesses a falling trajectory with her export of agricultural goods relative to aggregate export merchandise, much in congruence with the global scenario. This evokes the case that his plummeting presence of India's agriculture in global economic landscape is potentially attributable to sustained fall in demand for food and other traditional agricultural products. Particularly, the period onwards 2016 features a magnified form of such deteriorating trend in India's agricultural export, at the time when the existing regime on SPS standard gains momentum. Thus, erosion of export competitiveness of India's agriculture in face of rise in stringency of SPS is well evident. However, in the recent years there is witness of slight reversal of this dwindling situation but at global level much in contrast with India.

Coming to the context of how the observed dynamics in the global presence of India's agricultural sector, in terms of shares of agriculture and food in total export merchandise, as well as their year-on-year basis growth rate (as shown in Table 9.1), can be validated with the corresponding dynamics in her comparative advantage in the global agricultural trade (see Table 9.1). Following Balassa

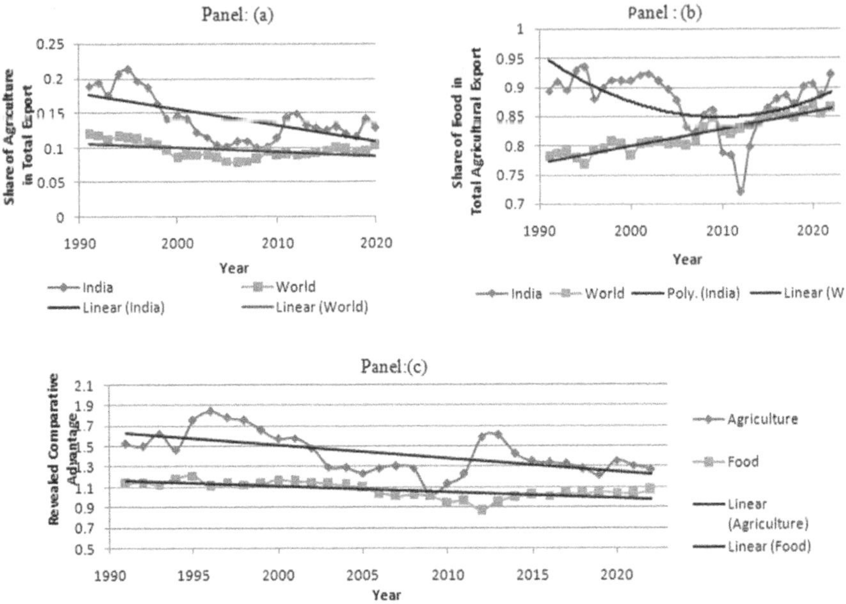

Fig. 9.2. India's Agriculture in Global Landscape Post-Economic Liberalization. *Source*: Authors' adoption and computation from WTO data.

Table 9.1. Revealed Comparative Advantage and Growth Rate of Export India's Agricultural Sector in Post-liberalization Era.

Year	Revealed Comparative Advantage Score		Growth Rate of Export	
	Agriculture	Food	Agriculture	Food
1991	1.523808	1.134852		
1992	1.501355	1.152962	9.37	11.46
1993	1.626741	1.130171	13.36	11.45
1994	1.470683	1.189504	5.57	9.55
1995	1.755237	1.213817	43.71	44.72
1996	1.839678	1.109095	11.36	4.75
1997	1.781836	1.131095	−2.51	−0.40
1998	1.75252	1.129964	−9.15	−7.80
1999	1.659054	1.135433	−6.42	−6.54
2000	1.571531	1.175825	1.99	1.94
2001	1.577576	1.160953	6.35	7.59

(*Continued*)

Table 9.1. (*Continued*)

Year	Revealed Comparative Advantage Score		Growth Rate of Export	
	Agriculture	Food	Agriculture	Food
2002	1.482797	1.152791	9.20	9.44
2003	1.301094	1.133652	4.25	2.90
2004	1.291475	1.123598	22.12	20.12
2005	1.237446	1.095111	16.76	14.15
2006	1.288068	1.043362	21.20	15.21
2007	1.30859	1.015995	30.91	29.30
2008	1.285296	1.028504	30.37	34.96
2009	1.039072	1.013817	−22.90	−22.06
2010	1.132944	0.957983	41.03	29.03
2011	1.234004	0.960225	49.27	48.41
2012	1.585613	0.870219	22.92	13.05
2013	1.611033	0.955732	10.75	22.58
2014	1.428927	1.002909	−8.64	−3.81
2015	1.351677	1.028405	−19.24	−16.80
2016	1.33879	1.020541	−4.56	−2.98
2017	1.326254	1.047112	17.45	18.28
2018	1.283258	1.034377	0.24	−1.48
2019	1.21723	1.055489	−4.00	−0.59
2020	1.361408	1.042567	4.02	4.53
2021	1.310997	1.036841	29.89	27.00
2022	1.272915	1.073418	8.31	12.81

Source: Authors' computation from WTO data.

(1956), degree of comparative advantage of a country in a commodity can be estimated from the ratio of country's share in the total import of the respective commodity by her trading partners to the share of the country in the total import of all the commodities made by her trading partners. Accordingly, India's comparative advantage in agriculture and food at the global level has been derived from India's share in world import of agriculture in ratio with her share in world import of all traded items. Thus, evident from trend lines in Panel 1(c) of Fig. 9.2, India has been posting an insipid picture in her test of competition edge in global supply chain of food and agricultural products. Moreover, as shown in Table 9.1, India's score on comparative advantage in food dropped below unity (which is considered as threshold for a significant comparative advantage) during the period from 2009 till 2014, and nonetheless, the recovery noticed afterwards could not restore her average score observed before 2009.

Conclusion

To sum up, for effective rehabilitation of India's glory in global trade in agriculture, and furthermore, to leverage her economic sovereignty there is pressing need for the restoration of indigenous form of agriculture, a kind of which is zero-budget natural farming. The capitalist mode of production is agricultural sector, which is emergent in LDCs including India, and is gaining further momentum with the inception of foreign capital inflow, in form of FDI, may held responsible for the increased dependence of market-based inputs, including chemical fertilizers and pesticides and even more, the high-yielding variety seeds (as replacement of traditional seeds, sourced internally from the production itself) and thus can be seen as challenge to the affordability of agriculture as livelihood practice, at large. Hence, for the sake of agriculture to evolve into an environmentally sustainable means of livelihood, there is immense importance in the adoption of natural farming and finding ways to foster its commercial prospect. In other words, ZBNF can assume remedial role against hazardous implication of inorganic farming for environment by reversing loss of biodiversity caused thereof. To this end the role of state-sponsored research towards the progress of indigenous technology to the state-of-art level is indispensable.

References

Agarwal, M. (1988). A comparative analysis of India's export performance, 1965-80. *Indian Economic Review*, July–December, *23*(2), 231–262.

Aggarwal, P. K., et al. (2018). The climate-smart village approach: Framework of an integrative strategy for scaling up adaptation options in agriculture. Ecology and Society, *23*(1), DOI: 10.5751/ES-09844-230114

Bargout, R. N. (2014). Ecological agriculture and sustainable adaptation to climate change: A practical and holistic strategy for Indian smallholders. *Consilience*, *9*(12), 95–122.

Cheong, D., Jansen, M., & Peters, R. (Eds.). (2013). Shared harvests: Agriculture, trade, and employment. In *International labour office and United Nations conference on trade and development*. ILO & UNCTAD.

Duddigan, S., Collins, C. D., Hussain, Z., Osbahr, H., Shaw, L. J., Sinclair, F., Sizmur, T., Thallam, V., & Ann Winowiecki, L. (2022). Impact of zero budget natural farming on crop yields in Andhra Pradesh, SE India. *Sustainability, 14*(3), 1689. https://doi.org/10.3390/su14031689

Duddigan, S., Shaw, L. J., Sizmur, T., Gogu, D., Hussain, Z., Jirra, K., Kaliki, H., Sanka, R., Sohail, M., Soma, R., Thallam, V., Vattikutti, H., & Collins, C. D. (2023). Natural farming improves crop yield in SE India when compared to conventional or organic systems by enhancing soil quality. *Agronomy for Sustainable Development*, *43*, 31. https://doi.org/10.1007/s13593-023-00884-x

Gill, S., & Brar, J. S. (1996). Global market and competitiveness of Indian agriculture: Some issues. *Economic and Political Weekly*, *31*(32), 2167–2177.

Gulati, A. (2002). Indian agriculture in a globalizing world. *American Journal of Agricultural Economics*, *84*(3), 754–761.

Khurana, A., & Kumar, V. (2020). *State of organic and natural farming: Challenges and possibilities*. Centre for Science and Environment. https://www.cseIndia.org/content/downloadreports/10346

Laishram, C., Vashishat, R. K., Sharma, S., Rajkumari, B., Mishra, N., Barwal, P., Vaidya, M. K., Sharma, R., Chandel, R. S., Chandel, A., Gupta, R. K., & Sharma, N. (2022). Impact of natural farming cropping system on rural households – Evidence from Solan District of Himachal Pradesh, India. *Frontiers in Sustainable Food Systems*, *6*, 878015. https:doi.org/10.3389/fsufs.2022.878015

Lee, D. R. (2005). Agricultural sustainability and technology adoption: Issues and policies for developing countries. *American Journal of Agricultural Economics*, *87*(5), 1325–1334.

Nayyar, D., & Sen, A. (1994). International trade and the agricultural sector in India. *Economic and Political Weekly*, *29*(20), 1187–1203.

NSSO. (2013). *All India debt and investment survey report*. NSSO, 70th round.

Sarah, D., Collins, C. D., Hussain, Z., Osbahr, H., Shaw, L. J., Sinclair, F., Sizmur, T., Thallam, V., & Winowiecki, L. A. (2022). Impact of zero budget natural farming on crop yields in Andhra Pradesh, SE India. *Sustainability*, 14(3), 1689.

Sharma, K., Ravisankar, N., Jain N. K., & Sarangi, S. K. (2023). Natural farming: Current status, research and case studies. *Indian Journal of Agronomy*, *68* (XXII Biennial National Symposium Special issue), S1–S15.

Tripathi, S., Shahidi, T., Nagbhushan, S., & Gupta, N. (2018, September). Zero budget natural farming for the sustainable development goals. Council on Energy, Environment and Water (CEEW), Andhra Pradesh, India.

Viswanathan, P. K., Thapa, G. B., Routray, J. K., & Ahmad, M. M. (2012). Agrarian transition and emerging challenges in Asian Agriculture: A critical assessment. *Economic and Political Weekly*, *47*(4), 41–50.

Vyas, V. S. (1999). Agricultural trade policy and export strategy. *Economic and Political Weekly*, *34*(13), A27–A33. WTO. (2023). *WTO statistical review*.

Section B

Environmental Insinuations of Agricultural Practices Across Countries and Groups

Chapter 10

Determining the Most Appropriate Strategy for Reducing Greenhouse Gas Emissions Caused by the Agriculture Sector Using DEMATEL

Serkan Eti[a], *İrfan Ersin*[b], *Yaşar Gökalp*[c], *Çağatay Çağlayan*[d] *and Duygu Yavuz*[d]

[a]*The Vocational School, İstanbul Medipol University, Turkey*
[b]*The Social Sciences Vocational School, İstanbul Medipol University, Turkey*
[c]*The School of Health, İstanbul Medipol University, Turkey*
[d]*The School of Business, İstanbul Medipol University, Turkey*

Abstract

Agriculture is an activity that plays an important role in human life. Similarly, the agricultural sector plays an important role in the national economy. One of the biggest problems of the agricultural sector is the carbon gas it produces during production. Fertilizing activities and tools used in plowing the fields cause this gas to be produced. The release of the said gas into nature causes serious damage to the environment. Therefore, carbon emissions in the agricultural sector are of vital importance. In line with this purpose, it is aimed to determine the most appropriate strategy for carbon emission in this study. As a result of the DEMATEL analysis, it was seen that the most appropriate strategy was effective regulations and auditing.

Keywords: Agriculture sector; CO_2 emissions; DEMATEL; carbon capture technology; renewable energy; effective regulations and auditing; energy efficiency; training

Sustainable Agricultural Practices: Economic and Environmental Implications, 147–159

doi:10.1108/978-1-83608-336-820241010

Introduction

Environmental pollution is a worldwide problem that endangers all forms of life. The rapid increase in economic and social activities also rapidly increases CO_2 emissions, which has the highest share of greenhouse gas emissions (Vermont & De Cara, 2010). Therefore, the environmental factor is exposed to problems such as global warming, climate change and pollution. With rising global demand for food, energy and other production-based sectors, it is becoming more important to set up systems with greater production capacity and to shift focus toward integrated production endeavors. Climate change and global warming are recognized as the most serious and critical environmental problems nowadays (Martínez et al., 2023). To tackle environmental issues including global warming and climate change, governments and policymakers work to limit CO_2 emissions. One of the direct ways to reduce carbon emissions could be to reduce fossil-based energy consumption (Ortiz, 2021). The manufacturing, industrial, tourist, transportation and agricultural sectors of both developed and developing nations' economies have relied heavily on fossil-based energy in recent years. Intensifying global temperatures due to climate change, which is mostly driven by the combustion of fossil fuels, may pose a few issues. In this setting, it's useful for a wide range of things, including water management, forest and plant preservation, biodiversity conservation, agricultural productivity and health promotion (Zornoza et al., 2016). Conversely, the negative consequences of climate change, such as shifts in precipitation and temperature, drought, desertification and natural catastrophes, have been growing in frequency and intensity in recent years. Because of the negative impact on agricultural output and growth rates, this is a serious issue for the economy. Possible outcomes of climate change are warmer temperatures, altered precipitation patterns and increased levels of atmospheric carbon dioxide (CO_2). With a growing global population comes a greater need for food, yet climate change is making it harder for agriculture (Agovino et al., 2019).

The agricultural sector's productivity decreases because of climate change-related changes in temperature and precipitation, which in turn affects the quantity and quality of goods. As a result, agriculture will have a smaller impact on GDP, which will slow the economy. It is important to work on solutions to adapt to climate change and mitigate its negative effects to head off potential problems in the future. To stop production losses in the agricultural sector, it is important to come up with new ways to farm to fight climate change and adapt to its effects (Lin & Xu, 2018). The conservation of arable land, the promotion of organic farming practices and the improvement of agricultural irrigation technology are all examples of sustainable agricultural policy. The use of renewable energy sources to replace fossil fuels and the shift toward better industrial practices are equally crucial to this cause (Hossain & Chen, 2022). Since the ecosystem is a whole, it is possible to prevent the harmful effects of climate change with measures to be taken not only at the regional but also at the international level (Yüksel et al., 2023; Kou et al., 2023). For this reason, it is an important step for all countries to support incentive policies for the reduction of fossil fuel use and the use of renewable energy sources. There are three ways in which the greenhouse effect affects farming: To begin, there may be a direct relationship between elevated CO_2 levels in the atmosphere and enhanced plant and weed development. Second, variations in temperature, precipitation and sunshine due to increased CO_2 may have an impact on agricultural output and animal welfare (Xu et al., 2023). Finally, sea-level rises can lead to loss of salinity of farmland and groundwater in coastal areas. In recent years, thanks to the increase in incentives available especially in the renewable energy sector and the development of interest in renewable energy all over the world; By-products and harmful substances resulting from energy production have started to decrease and awareness has been created on this issue (Waheed et al., 2018). Detailed information on the most common types of renewable energy used in agricultural settings is provided:

i) Solar energy: It has a wide range of agricultural applications, including greenhouse heating and cooling, greenhouse lighting, agricultural product drying and field irrigation.
ii) Grain dust, wheat straw and nutshells, together with advanced biofuels like bioethanol and biogas, all provide opportunities for the development of new forms of energy production.
iii) Geothermal energy: It may be put to work warming the ground and drying crops in a wide variety of settings including fishing, barns, soil improvement, greenhouses and open fields.
iv) Generating electrical energy: Wind power may be used to power machinery that both drenches fields and grinds certain crops.
v) Hydroelectric: Its multipurpose nature makes it ideal for power plants, water treatment plants and municipal water systems, and it also helps farmers more easily divide up water resources.

In light of the findings, governments should increase investments in renewable energy resources and create policies that will encourage the private sector for more investment in order to effectively combat global warming and climate change problems and reduce foreign dependency on energy in line with sustainable development goals. Much can be done to reduce global warming, such as using animal manure in the agricultural sector, converting agricultural products and residues into energy and using renewable energy in agricultural activities. The government enacts ecologically friendly agriculture regulations to mitigate the industry's growing contribution to climate change (Li et al., 2016). Farmers should be guided and encouraged to benefit from solar, geothermal and wind energy in agricultural activities. In addition, in order to reduce the negative effects of urbanization on the environment, environmentally friendly urbanization plans should be prepared, and policies should be followed to encourage renewable energy consumption in cities because it is likely that climate change will adversely affect food production (Anwar et al., 2013). Considering the importance of these activities in terms of both the continuation of human generation and the sustainability of the economy, the necessity of taking precautions against climate change emerges. In short, the agricultural sector is the sector mostly affected by climate change. However, the effects of climatic effects on agriculture differ worldwide.

It is not entirely possible to determine how climate change will affect agriculture. Climate change and the resulting rise in CO_2 levels are predicted to have profound consequences on agriculture across the world, particularly in tropical areas, due to shifts in average temperature and precipitation (Luo et al., 2017). Changes in temperature, weather events and pest and disease patterns may also affect agricultural output. In addition, agricultural lands suitable for the cultivation of basic plants may also undergo geographical changes due to climate change. For this reason, it is crucial to have good policies in place to ensure agriculture and the economy can continue to thrive over the long haul.

Literature Review

The most serious environmental problems of our age are global warming, climate change and CO_2 emissions. Many sectors such as construction, logistics, livestock, fisheries, health, tourism and education are affected by these problems. Another important sector affected is agriculture. The agricultural sector is a sector that directly affects both people's lives and the development of countries. For this reason, countries are working to reduce the CO_2 emission rate in the agricultural sector. One of the things that needs to be done to reduce CO_2 emissions is the use of carbon capture technology. González-Sánchez et al. (2012) and Ghiat et al. (2021) examined the relationship between carbon capture technology and agriculture in their studies. Accordingly, it has been stated that the use of carbon capture technology is an effective way to reduce CO_2 emissions. Mikulcic et al. (2019) and Pant et al. (2021) carried out a study to reduce fossil fuel consumption, which causes CO_2 emissions. These studies suggesting the use of renewable energy instead of fossil fuels stated that carbon capture technology is an important way to reduce the effects of CO_2 emissions. Rubin et al. (2012) conducted a study on the development of carbon capture technology in the process of reducing CO_2 emissions. Expressing that one of the biggest discussion topics for carbon capture technology is cost, Rubin et al. (2012) state that carbon capture and storage technologies are important tools to reduce carbon emissions. Markewitz et al. (2019) state that the cement industry is at the forefront of the sectors that cause carbon emissions. Therefore, it carried out a study to reduce carbon emissions for the cement industry in Germany. The results of the study underline that carbon capture technology has an important role in reducing carbon emissions.

CO_2 emissions can harm people and the environment. Many ways can be followed to reduce this emission. Increasing the use of renewable energy is one of these ways. Waheed et al. (2018) conducted a study to examine the effects on renewable energy consumption, agricultural production and CO_2 emissions in Pakistan. Autoregressive distributed latency analysis was used in a study conducted using data from 1990 to 2014. According to the results of the study, it was determined that the use of renewable energy in the long term reduces CO_2 emissions. Rahman et al. (2022) and Liu et al. (2017) conducted a study on what needs to be done to strengthen agriculture. As a result of these studies, which have been analyzed with different methods, they argue that more efficient agriculture can be made by preventing the use of renewable energy from CO_2 emissions. Zheng et al. (2021) conducted a study to reduce carbon emissions in China. In the study, regression analysis was performed on the data of the years 2008–2017. In the study, it is stated that renewable energy is a hindering force on carbon emission intensity. It is also stated that a 1% increase in renewable energy development causes a decrease in carbon emission intensity in the range of 0.028–0.043%. Jin and Kim (2018) conducted a study that included a comparison of renewable energy and nuclear energy to reduce carbon emissions. Panel cointegration analysis and Granger causality analysis were applied in the study using the data of 30 countries for the period 1990–2014. The results of the study show that the use and dissemination of renewable energy are beneficial both economically and ecologically.

Another application that can reduce CO_2 emissions is legal regulations. Thanks to the measures to be taken with legal regulations, it will be possible to minimize the negative impact of CO_2 emissions on agriculture. Chen et al. (2017) and Kalphle et al. (2018) state that environmental pollution affects agricultural production and yield in their studies. They emphasize that it is important to ensure productivity in the agricultural sector by reducing this pollution. They underline that an important way of reducing environmental pollution is legal regulations. Similarly, Gardner (2019) examined the competitiveness issues for US agricultural activities in his study for the United States of America (USA). In his study, Gardner states that farm and agricultural practices can cause many environmental problems. He says that the solution to this situation can be realized with the necessary regulations and measures. Yin et al. (2022) examined the effect of environmental regulations on carbon emissions in their study in China. In this study, recommendations are made for the Chinese state, which is committed to achieving carbon neutrality by 2060. Accordingly, it has been concluded that environmental regulations should be carried out effectively to increase the use of clean energy and reduce carbon emissions. Ulucak et al. (2020) examined the relationship between environmental regulations and reduction of carbon emissions for BRICS countries. Econometric analysis methods were used in the study, in which data from 1995 to 2016 were used. It was concluded that for BRICS countries, climate change and carbon emission levels are related to environmental regulations. Gu (2022) examined the relationship of landscaping policies with energy saving and carbon emission reduction in cities in China. As a result of the study, it was stated that environmental regulation policies contributed significantly to energy savings and reduction of carbon emissions.

Ensuring energy efficiency is another application that can reduce CO_2 emissions. Energy efficiency means reducing the energy required to provide a product or service. Reducing the amount of energy consumption means less carbon emissions. Wang and Wu (2021) conducted a study on carbon emissions and energy efficiency in China. In this study using data envelopment analysis, regional energy efficiency in China was evaluated in the context of carbon neutrality. According to the results of the study, CO_2 emissions are lower in regions with high energy efficiency. Hassan et al. (2022) and Wang and Wu (2020) examined the data of OECD countries and China separately in their study on reducing carbon emissions by providing energy efficiency. Studies argue that urgent global action is needed to decarbonize the energy sector. Accordingly, they state that providing energy efficiency reduces CO_2 emissions. Kairies-Alvarado et al. (2021), in their study in public buildings in Chile, examined the contribution of energy efficiency to the reduction of carbon emissions. In the study, it was concluded that providing energy efficiency can reduce carbon emissions. Huang and Wu (2021) examined the relationship between energy efficiency and carbon emission reduction in the cement industry. In the study conducted in Taiwan, it was concluded that energy efficiency and energy saving are important factors in reducing carbon emissions. Alamoush et al. (2020) comparatively examined the impact of energy efficiency on carbon emissions for developed and developing countries. The study was carried out by considering the data of 28 developed and 34 developing countries between the years 1990–2017. A 1% increase in energy efficiency in developed countries causes a 1.24% decrease in carbon emissions. According to another result, a 1% increase in energy efficiency in developing countries causes a 1.19% decrease in carbon emissions.

According to studies in literature, one of the practices that can reduce CO_2 emissions is education. Thanks to the training on CO_2 emission to be given to users, the damages caused by carbon emissions will be reduced. With these trainings, the awareness level of the users will also be increased. Zafar et al. (2021) examined the relationship between education, carbon emissions and several different variables for 22 different countries. The regression method was used in this study, which was conducted using the data of the years 1986–2017. The results of the study have findings that increasing the level of education reduces carbon emissions. Zafar et al. (2020) carried out a study to measure renewable energy consumption, education and their contribution to environmental quality in OECD countries. Data for the years 1990–2015 were used in the study. According to the results of the study, it was concluded that education reduces CO_2 emissions. Adikari et al. (2023) state in their study that many environmental problems are caused by human beings and carbon emission is one of these problems. In this study, in which Sri Lanka's 1978–2019 data were used, time series analysis was used. According to the results of the study, it is underlined that carbon emissions can be reduced by investing more in education. Liu et al. (2022) conducted a study on the relationship between reducing carbon emissions and tourism. The study was conducted using data from 70 countries for the years 2000–2017. It has been emphasized that education has a significant impact on reducing carbon emissions.

As a result of the literature study, the following conclusions were reached:

i. Agriculture is of great importance for people to survive.
ii. However, one of the obstacles to achieving agricultural productivity is CO_2 emissions.
iii. Many studies have been carried out in the literature to reduce CO_2 emissions.
iv. As a result of the studies, the use of carbon capture technology, the use of renewable energy, legal regulations, providing energy efficiency and providing the necessary training draw attention as the issues that can reduce CO_2 emissions.
v. It is not possible to apply all of these aspects at the same time. Therefore, it is important to rank these criteria in order of importance.

In this context, the aim of this study is to develop a strategy to minimize the effect of CO_2 emissions in the agricultural sector. With this study, it will be possible to achieve what needs to be done to minimize CO_2 emissions.

Dematel Method

The DEMATEL method (Decision-Making Trial and Evaluation Laboratory) is a multi-criteria decision-making method used to analyze complex systems and understand the interrelationships between criteria that influence a particular problem (Eti et al., 2022). The method was first introduced in 1972 by the Japanese researcher and engineer, Kazuo Inoue.

The DEMATEL method is used for various purposes. It can be summarized as

- Problem-solving: The method helps to identify the root causes of a problem and determine the relationships between different criteria affecting the problem (Baş et al., 2022; Eti et al., 2022).
- Decision-making: The method helps decision-makers to understand the complex interrelationships between various criteria affecting a particular decision and identify the most critical criteria that need to be addressed (Eti et al., 2023; Yüksel et al., 2022a).
- Strategic planning: The method helps to identify the key criteria that need to be addressed to achieve the strategic objectives of an organization (Yüksel et al., 2022a, 2022b).

The DEMATEL method involves creating a matrix of cause-and-effect relationships between different criteria using expert opinions and data. Calculate the total impact matrix with a series of operations. The matrix is then transformed into a digraph, which allows for a visualization of the interdependencies and feedback loops between different criteria. This helps to identify the most influential criteria and determine the direction of causality. Finally, the method helps to prioritize the criteria based on their importance and the level of control that can be exercised over them (Kalkavan et al., 2021).

The DEMATEL method offers several advantages. It can be summarized as

1. Comprehensive analysis: The DEMATEL method allows for a comprehensive analysis of complex systems and helps to identify the interdependencies and feedback loops between different criteria. This helps to provide a more complete picture of the problem or decision at hand (Baş et al., 2022; Song et al., 2020).
2. Expert opinions: The method relies on expert opinions and knowledge, which allows for a more nuanced and informed analysis (Yüksel et al., 2022a).
3. Visual representation: The method produces a visual representation of the interrelationships between different criteria, which makes it easier to communicate the results and understand the problem or decision (Yüksel et al., 2022b).
4. Prioritization: The method helps to prioritize the most critical criteria based on their importance and the level of control that can be exercised over them. This allows for a more targeted approach to problem-solving and decision-making.
5. Flexibility: The DEMATEL method is flexible and can be adapted to different types of problems and decision-making situations (Liu et al., 2021).

6. Quantitative and qualitative analysis: The method allows for both quantitative and qualitative analysis that provides a more complete and nuanced understanding of the problem or decision at hand (Acuña-Carvajal et al., 2019; Bhuiyan et al., 2022).
7. Validation: The DEMATEL method allows for the validation of the results through sensitivity analysis and testing of different scenarios (Bhuiyan et al., 2022; Dinçer et al., 2019).

Overall, the DEMATEL method is a useful tool for decision-makers and analysts looking to gain a comprehensive understanding of complex systems and prioritize the most critical criteria for action (Eti et al., 2023; Özdemirci et al., 2023).

Application

It is aimed to determine the most appropriate strategy to reduce the impact of CO_2 emissions in the agricultural sector. For this purpose, DEMATEL method was used. Strategies to reduce CO_2 emissions have been determined based on the literature: Carbon Capture Technology (C1), Renewable Energy (C2), Effective Regulations and Auditing (C3), Energy Efficiency (C4) and Training (C5). Opinions of three experts were taken to determine which of these strategies was the most appropriate. The experts in question were selected from people who are experienced in their field and have academic studies. The decision matrix was formed by the average of the evaluations of these experts. The decision matrix created is given in Panel A of Table 10.1.

Table 10.1. Decision Matrix.

Panel A: Decision Matrix					
	C1	**C2**	**C3**	**C4**	**C5**
C1	0.000	2.667	2.667	2.333	2.000
C2	2.667	0.000	2.333	2.000	2.667
C3	4.000	3.667	0.000	4.000	3.333
C4	2.000	2.333	2.333	0.000	2.333
C5	2.667	2.333	3.000	2.333	0.000
Panel B: Total Effect Matrix					
	C1	**C2**	**C3**	**C4**	**C5**
C1	0.365	0.508	0.488	0.481	0.455
C2	0.514	0.354	0.471	0.462	0.486
C3	0.733	0.706	0.482	0.706	0.662
C4	0.460	0.468	0.451	0.324	0.451
C5	0.541	0.515	0.527	0.505	0.359

(Continued)

Table 10.1. (*Continued*)

Panel C: Importance Level of Strategies

	D	*R*	*D* + *R*	*W*
Carbon capture technology	2.297	2.613	4.910	0.197
Renewable energy	2.286	2.550	4.836	0.194
Effective regulations and auditing	3.289	2.418	5.708	0.229
Energy efficiency	2.154	2.478	4.632	0.186
Training	2.446	2.413	4.859	0.195

Source: Authors' computations.

DEMATEL operations are applied over the decision matrix. First, the normalization and then the total effect matrix (*T*) were obtained, respectively. The calculated *T* matrix is given in Panel B of Table 10.1. D and R matrices were obtained by summing the row and column elements of the total effect matrix. By adding the calculated D and R matrices, the importance levels of the strategies were determined. The results of the mentioned transactions and the importance degrees of the strategies are given in Panel C of Table 10.1.

When the results of the analysis are examined, it is seen that the most important strategy in CO_2 emission in the agricultural sector is 'Effective Regulations and Auditing'. While the 'Carbon Capture Technology' strategy took the second place, 'Energy Efficiency' took the last place.

Conclusion

Climate change will unavoidably have some regional and global repercussions. It is expected that the impacts of global climate change on agriculture, forestry and vegetation, clean water supplies, sea level, energy, human health and biodiversity will become apparent. The increasing impact of climate change, which is one of the biggest problems encountered on a global scale, directs researchers to work to reduce the destructive effects of climate change in environmental, economic dimensions and social. There are some chain effects of global climate change on economic life and social (Moiseev et al., 2023; Wan et al., 2023). Climate change, which has changed as a result of the increase in the world population and the increase in human welfare, has become a threat because of production and food security. Today, due to the impossibility of fully compensating for the effects of global greenhouse gas emissions, there is a greater emphasis on mitigating climate change and modifying policy. According to the Fourth Assessment Report of the Intergovernmental Panel on Climate Change (IPCC), adaptation studies are crucial for mitigating the effects of climate change.

Just as productivity and production in the agricultural sector are affected by climate change, fossil fuel tools and equipment used in agricultural production can also contribute to climate change. To reduce greenhouse gas emissions and sources of climate change, it is important to determine the most appropriate strategy in the agricultural sector. This study aimed to determine the most appropriate strategy to reduce the impact of CO_2 emissions in the agricultural sector, and for this purpose, the DEMATEL method, in which expert opinions were evaluated, was used. Strategies to reduce CO_2 emissions have been determined based on literature. These criteria are included in the analysis as Carbon Capture Technology (C1), Renewable Energy (C2), Effective Regulations and Control (C3), Energy Efficiency (C4) and Education (C5). As a result of the analysis, it is seen that the most important strategy in CO_2 emissions in the agricultural sector is 'Effective Regulations and Control'. 'Carbon Capture Technology' took second place, while 'Energy Efficiency' took last place.

The agricultural industry must invest in the research and development of innovative agricultural practices to battle and adapt to the effects of climate change. This will help the industry avoid production losses. In this context, it is seen as essential to take actions such as the research and development of agricultural irrigation technologies, the promotion of organic farming practices, the preservation of fertile agricultural lands, the establishment of investment and legal frameworks for these actions and the expansion of agricultural inspections. Such policies will contribute to sustainable economic development. Using carbon capture technologies has also been identified as an important strategy to reduce greenhouse gas emissions. Reparative agriculture is considered among the carbon capture technologies; 55% of the organic matter in the soil consists of carbon. Therefore, increasing the organic matter ratio means drawing CO_2 directly from the atmosphere and storing it in the soil. In 1-m-deep soil, increasing the organic matter rate by 1% is synonymous with creating 14 kg of organic matter per square meter. As a result, 28 kg of CO_2 is buried in the soil. Thus, reparative agricultural practices; it is an important tool in achieving the goal of improving the soil ecosystem, sustainable food production and keeping global warming below 2°C (Akhuy, 2020). Finally, for energy efficiency, reducing the number of vehicles used with fossil fuels in agricultural production, encouraging the use of vehicles based on renewable energy and raising awareness of farmers for energy saving are seen as important policies (Ai et al., 2023; Mikhaylov et al., 2023). All policies will have effective results in both reducing climate change and reducing greenhouse gas emissions.

References

Acuña-Carvajal, F., Pinto-Tarazona, L., López-Ospina, H., Barros-Castro, R., Quezada, L., & Palacio, K. (2019). An integrated method to plan, structure and validate a business strategy using fuzzy DEMATEL and the balanced scorecard. *Expert Systems with Applications*, *122*, 351–368.

Adikari, A. P., Liu, H., Dissanayake, D. M. S. L. B., & Ranagalage, M. (2023). Human capital and carbon emissions: The way forward reducing environmental degradation. *Sustainability*, *15*(4), 2926.

Agovino, M., Casaccia, M., Ciommi, M., Ferrara, M., & Marchesano, K. (2019). Agriculture, climate change and sustainability: The case of EU-28. *Ecological Indicators*, *105*, 525–543.

Ai, R., Zheng, Y., Yüksel, S., & Dinçer, H. (2023). Investigating the components of fintech ecosystem for distributed energy investments with an integrated quantum spherical decision support system. *Financial Innovation*, *9*(1), 27.

Akhuy, M. (2020). Ekosistem Sağlığı İçin Onarıcı Tarım. Retrieved February 23, 2023, from https://www.bugday.org/blog/ekosistem-sagligi-icin-onarici-tarim/

Alamoush, A. S., Ballini, F., & Ölçer, A. I. (2020). Ports' technical and operational measures to reduce greenhouse gas emission and improve energy efficiency: A review. *Marine Pollution Bulletin*, *160*, 111508.

Anwar, M. R., Liu, D. L., Macadam, I., & Kelly, G. (2013). Adapting agriculture to climate change: A review. *Theoretical and Applied Climatology*, *113*, 225–245.

Baş, H., Eti, S., & Ersin, I. (2022). Importance of renewable energy investments for qualified workforce. In H. Dinçer & S. Yüksel (Ed.), *Multidimensional strategic outlook on global competitive energy economics and finance* (pp. 141–152). Emerald Publishing Limited. https://doi.org/10.1108/978-1-80117-898-320221015

Bhuiyan, M. A., Dinçer, H., Yüksel, S., Mikhaylov, A., Danish, M. S. S., Pinter, G., Uyeh, D. D., & Stepanova, D. (2022). Economic indicators and bioenergy supply in developed economies: QROF-DEMATEL and random forest models. *Energy Reports*, *8*, 561–570.

Chen, Y. H., Wen, X. W., Wang, B., & Nie, P. Y. (2017). Agricultural pollution and regulation: How to subsidize agriculture? *Journal of Cleaner Production*, *164*, 258–264.

Dinçer, H., Yüksel, S., & Martínez, L. (2019). Interval type 2-based hybrid fuzzy evaluation of financial services in E7 economies with DEMATEL-ANP and MOORA methods. *Applied Soft Computing*, *79*, 186–202.

Eti, S., Dinçer, H., Gökalp, Y., Yüksel, S., & Kararoğlu, D. (2023). Identifying key issues to handle the inflation problem in the healthcare industry caused by energy prices: An evaluation with decision-making models. In *Managing inflation and supply chain disruptions in the global economy* (pp. 162–178).

Eti, S., Dinçer, H., Yüksel, S., & Gökalp, Y. Analysis of the suitability of the solar panels for hospitals: A new fuzzy decision-making model proposal with the T-spherical TOP-DEMATEL method. *Journal of Intelligent & Fuzzy Systems* (Preprint), 1–13.

Gardner, B. L. (2019). Environmental regulation and the competitiveness of US agriculture. In M. E. Bredahl, N. Ballenger, J. C. Dunmore, & T. L. Roe (Eds.), *Agriculture, trade, and the environment: Discovering and measuring critical linkage* (pp. 215–230). Westview Press.

Ghiat, I., Mahmood, F., Govindan, R., & Al-Ansari, T. (2021). CO_2 utilisation in agricultural greenhouses: A novel 'plant to plant' approach driven by bioenergy with carbon capture systems within the energy, water and food Nexus. *Energy Conversion and Management*, *228*, 113668.

González-Sánchez, E. J., Ordóñez-Fernández, R., Carbonell-Bojollo, R., Veroz-González, O., & Gil-Ribes, J. A. (2012). Meta-analysis on atmospheric carbon capture in Spain through the use of conservation agriculture. *Soil and Tillage Research*, *122*, 52–60.

Hassan, T., Khan, Y., He, C., Chen, J., Alsagr, N., & Song, H. (2022). Environmental regulations, political risk and consumption-based carbon emissions: Evidence from OECD economies. *Journal of Environmental Management*, *320*, 115893.

Hossain, M. A., & Chen, S. (2022). The decoupling study of agricultural energy-driven CO_2 emissions from agricultural sector development. *International Journal of Environmental Science and Technology*, *19*(5), 4509–4524.

Huang, Y. H., & Wu, J. H. (2021). Bottom-up analysis of energy efficiency improvement and CO2 emission reduction potentials in the cement industry for energy transition: An application of extended marginal abatement cost curves. *Journal of Cleaner Production*, *296*, 126619.

Jin, T., & Kim, J. (2018). What is better for mitigating carbon emissions – Renewable energy or nuclear energy? A panel data analysis. *Renewable and Sustainable Energy Reviews*, *91*, 464–471.

Kairies-Alvarado, D., Muñoz-Sanguinetti, C., & Martínez-Rocamora, A. (2021). Contribution of energy efficiency standards to life-cycle carbon footprint reduction in public buildings in Chile. *Energy and Buildings*, *236*, 110797.

Kalkavan, H., Baş, H., Ersin, İ., Eti, S., & Yüksel, S. (2021). Defining appropriate government strategies to reduce unemployment during COVID-19 pandemics. In H. Dincer & S. Yüksel (Eds.), *Management strategies to survive in a competitive environment. Contributions to management science*. Springer. https://doi.org/10.1007/978-3-030-72288-3_11

Kaphle, A., Navya, P. N., Umapathi, A., & Daima, H. K. (2018). Nanomaterials for agriculture, food and environment: Applications, toxicity and regulation. *Environmental Chemistry Letters*, *16*, 43–58.

Kou, G., Dincer, H., Yuksel, S., & Alotaibi, F. S. (2023). Imputed expert decision recommendation system for QFD-based omnichannel strategy selection for financial services. *International Journal of Information Technology & Decision Making*, *23*(1), 141–170.

Li, T., Baležentis, T., Makutėnienė, D., Streimikiene, D., & Kriščiukaitienė, I. (2016). Energy-related CO_2 emission in European Union agriculture: Driving forces and possibilities for reduction. *Applied Energy*, *180*, 682–694.

Lin, B., & Xu, B. (2018). Factors affecting CO_2 emissions in China's agriculture sector: A quantile regression. *Renewable and Sustainable Energy Reviews*, *94*, 15–27.

Liu, W., Dinçer, H., Eti, S., & Yüksel, S. (2021). Gaussian-based hybrid approach to entropy for analyzing energy efficiency of emerging economies. *Energy Reports*, *7*, 2501–2511.

Liu, X., Zhang, S., & Bae, J. (2017). The nexus of renewable energy-agriculture-environment in BRICS. *Applied Energy*, *204*, 489–496.

Liu, Z., Lan, J., Chien, F., Sadiq, M., & Nawaz, M. A. (2022). Role of tourism development in environmental degradation: A step towards emission reduction. *Journal of Environmental Management*, *303*, 114078.

Luo, Y., Long, X., Wu, C., & Zhang, J. (2017). Decoupling CO_2 emissions from economic growth in agricultural sector across 30 Chinese provinces from 1997 to 2014. *Journal of Cleaner Production*, *159*, 220–228.

Markewitz, P., Zhao, L., Ryssel, M., Moumin, G., Wang, Y., Sattler, C., Robinius, M., & Stolten, D. (2019). Carbon capture for CO2 emission reduction in the cement industry in Germany. *Energies*, *12*(12), 2432.

Martínez, L., Dinçer, H., & Yüksel, S. (2023). A hybrid decision making approach for new service development process of renewable energy investment. *Applied Soft Computing*, *133*, 109897.

Mikhaylov, A., Dinçer, H., & Yüksel, S. (2023). Analysis of financial development and open innovation oriented fintech potential for emerging economies using an integrated decision-making approach of MF-X-DMA and golden cut bipolar q-ROFSs. *Financial Innovation*, *9*(1), 1–34.

Mikulčić, H., Skov, I. R., Dominković, D. F., Alwi, S. R. W., Manan, Z. A., Tan, R., Duić, N., Mohamad, S. N. H., & Wang, X. (2019). Flexible carbon capture and utilization technologies in future energy systems and the utilization pathways of captured CO_2. *Renewable and Sustainable Energy Reviews*, *114*, 109338.

Moiseev, N., Mikhaylov, A., Dinçer, H., & Yüksel, S. (2023). Market capitalization shock effects on open innovation models in e-commerce: Golden cut q-rung orthopair fuzzy multicriteria decision-making analysis. *Financial Innovation*, *9*(1), 1–25.

Ortiz, A. M. D., Outhwaite, C. L., Dalin, C., & Newbold, T. (2021). A review of the interactions between biodiversity, agriculture, climate change, and international trade: Research and policy priorities. *One Earth*, *4*(1), 88–101.

Özdemirci, F., Yüksel, S., Dinçer, H., & Eti, S. (2023). An assessment of alternative social banking systems using T-spherical fuzzy TOP-DEMATEL approach. *Decision Analytics Journal*, 100184. https://doi.org/10.1016/j.dajour.2023.100184

Pant, D., Nadda, A. K., Pant, K. K., & Agarwal, A. K. (2021). *Advances in carbon capture and utilization* (pp. 3–7). Springer Singapore.

Rahman, M. M., Khan, I., Field, D. L., Techato, K., & Alameh, K. (2022). Powering agriculture: Present status, future potential, and challenges of renewable energy applications. *Renewable Energy*, *188*, 731–749.

Rubin, E. S., Mantripragada, H., Marks, A., Versteeg, P., & Kitchin, J. (2012). The outlook for improved carbon capture technology. *Progress in Energy and Combustion Science*, *38*(5), 630–671.

Song, W., Zhu, Y., & Zhao, Q. (2020). Analyzing barriers for adopting sustainable online consumption: A rough hierarchical DEMATEL method. *Computers & Industrial Engineering*, *140*, 106279.

Ulucak, R. (2020). How do environmental technologies affect green growth? Evidence from BRICS economies. *Science of the Total Environment*, *712*, 136504.

Vermont, B., & De Cara, S. (2010). How costly is mitigation of non-CO_2 greenhouse gas emissions from agriculture? A meta-analysis. *Ecological Economics*, *69*(7), 1373–1386.

Waheed, R., Chang, D., Sarwar, S., & Chen, W. (2018). Forest, agriculture, renewable energy, and CO_2 emission. *Journal of Cleaner Production*, *172*, 4231–4238.

Wan, Q., Miao, X., Wang, C., Dinçer, H., & Yüksel, S. (2023). A hybrid decision support system with golden cut and bipolar q-ROFSs for evaluating the risk-based strategic priorities of fintech lending for clean energy projects. *Financial Innovation*, *9*(1), 1–25.

Wang, Z., & Wu, Q. (2021). Carbon emission reduction and product collection decisions in the closed-loop supply chain with cap-and-trade regulation. *International Journal of Production Research*, *59*(14), 4359–4383.

Xu, X., Yüksel, S., & Dinçer, H. (2023). An integrated decision-making approach with golden cut and bipolar Q-ROFSs to renewable energy storage investments. *International Journal of Fuzzy Systems*, *25*(1), 168–181.

Yin, K., Liu, L., & Gu, H. (2022). Green paradox or forced emission reduction – The dual effects of environmental regulation on carbon emissions. *International Journal of Environmental Research and Public Health*, *19*(17), 11058.

Yüksel, S., Dinçer, H., Çağlayan, Ç., & Uluer, G. S. (2023). The role of nuclear energy to reduce carbon emission. In M. K. Pal (Ed.), *The impact of environmental emissions and aggregate economic activity on industry: Theoretical and empirical perspectives* (pp. 67–77). Emerald Publishing Limited.

Yüksel, S., Dinçer, H., Eti, S., & Adalı, Z. (2022a). Strategy improvements to minimize the drawbacks of geothermal investments by using spherical fuzzy modelling. *International Journal of Energy Research*, *46*(8), 10796–10807.

Yüksel, S., Dinçer, H., Mikhaylov, A., Adalı, Z., & Eti, S. (2022b). Key issues for the improvements of shallow geothermal investments. In *Sustainability in energy business and finance: Approaches and developments in the energy market* (pp. 183–194). Springer International Publishing.

Zafar, M. W., Saleem, M. M., Destek, M. A., & Caglar, A. E. (2022). The dynamic linkage between remittances, export diversification, education, renewable energy consumption, economic growth, and CO2 emissions in top remittance-receiving countries. *Sustainable Development*, *30*(1), 165–175.

Zheng, H., Song, M., & Shen, Z. (2021). The evolution of renewable energy and its impact on carbon reduction in China. *Energy*, *237*, 121639.

Zornoza, R., Rosales, R. M., Acosta, J. A., De La Rosa, J. M., Arcenegui, V., Faz, Á., & Pérez-Pastor, A. (2016). Efficient irrigation management can contribute to reduce soil CO_2 emissions in agriculture. *Geoderma*, *263*, 70–77.

Chapter 11

Sustainable Agricultural Practices: Economic and Environmental Implications

Subhabrata Ghosh[a] *and Krishna Singh*[b]

[a]*Department of Botany, Mahadevananda Mahavidyalaya, Barrackpore, India*
[b]*Corresponding author; Department of Economics, University of Gour, Banga, India*

Abstract

Sustainability promotes specific and balanced circumstances which favours the survival of human population with the surrounding Mother Nature with proper synchronization. This practice ensures the use of all available natural resources in an equilibrated manner to satisfy the present needs along with the requirements of the succeeding generations. Agricultural sustainability reckons the role of healthy environment; society and economy in an integrated manner to promote food security and socio-economic development of the rural individuals. Normal agricultural practices have an unfathomed burden on the environment. This leads to various forms of environmental degradation like air and water pollution, soil depletion, climate change and loss of biodiversity. The sustainable agricultural practices aim to protect the environment, expand the Earth's natural resource base and preserve or improve fertility of soil. Sustainable agriculture comprises with environmentally friendly farming methods that allow crops to be produced without harming natural systems. This prevents adverse impacts on soil, water, biodiversity as well as surrounding or downstream resources. In this present article we will analyse all issues related to mitigating environmental degradation through agricultural sustainability. Along with the proper assessment for agricultural sustainability is very crucial to understand the real scenario. This helps to take necessary measures for

Sustainable Agricultural Practices: Economic and Environmental Implications, 161–175

doi:10.1108/978-1-83608-336-820241011

designing and its actual execution. The study constructed a composite index of agricultural sustainability by applying the principal component analysis method. Significant differences in agricultural sustainability were found among the states in India. It is important to mention that agricultural sustainability improved for all states in 2019 compared to 2016.

Keywords: Agricultural practices; sustainability; composite index; principal component; food security; socio-economic development

Introduction

Sustainable agriculture utilizes environmental resources to encourage food production in connection with minimum impingement on the environment. Sustainability works along with environment through preserving its gradients in a suitable and most beneficial way to maintain the ecological constancy (Francis, 1990; Smith & Serna, 2023). This particular approach furnishes itself as a substitute for conventional agricultural practices with other food choices. This includes use of plant-based protein sources to promote environmental balance, a better health condition with less inflammation and overall growth. Sustainable agriculture employs all available local wealth that reduce water utilization. This approach works utterly with natural fertilizers that can maintain soil characters and create a balanced and worthy natural source for the succeeding generations.

The affinity between agricultural practices and the surrounding environment is composite in nature and it constitutes its multifarious existence. This relationship conveys some confirming and disconfirming attributes.

Positive Outlooks

Ecosystem Services

These are the welfares offered by the surrounding environments to the human populations for their existence (Costanza et al., 1997; Close & Australia. Department of The Environment, Water, Heritage, and the Arts, 2010). These services are of four types namely provisioning services, regulating services, supporting services and cultural services. Among this the provisioning services play the indispensable role for the overall ecesises of the ecosystem. These are water, raw materials and food sources. Regulating services describe its modulating properties by which it can influence climate, overall water cycle, etc. Supporting services depict its purpose in overall diversity manifestation, nutrient mobilization, etc. A cultural service explains its recreational and aesthetic attributes.

Admitting and perceiving ecosystem services is very important for sustainability. Various anthropogenic activities, like clearing of land for intensive agriculture and industrial purposes, can lead to pollution and global warming and can dismantle the efficiencies of these services.

Carbon Sequestration

Carbon sequestration has an essential role in extenuating climatic alteration by trapping and storing carbon dioxide from the air. This is the fundamental part of the global carbon cycle mechanism, involving storing carbon in oceans, forests, soil and geological establishments. Usually, carbon sequestration operation can be typified into three categories namely soil, biological and geological. Collectively through these approaches the overall crop production can be increased by sustainable way (Kumara et al., 2023; Sanaullah et al., 2019).

Negative Outlooks

Deforestation

Deforestation is the panoptic elimination of forests or trees, turning forested zones into non-forest land for various manmade purposes like farming, logging, urban development, or building infrastructure, etc. This malpractice can lead to interruption in the water cycle, change in the climatic pattern, overall biodiversity modification and all these consequences together result into changes in sociocultural status negatively.

Pesticide and Fertilizer Use

Indiscriminate use of chemical substances as pesticides and fertilizers can contribute pollution and associated with various alterations in the ecosystem functionality. The most dramatic effects are water pollution, increase in greenhouse gas content, biodiversity loss, inclusion of chemical substances in food, health concerns for agricultural workers and ultimate perturbation in the ecological stability.

Water Use

Agriculture is a major consumer of water, and unlawful water use or wrong irrigation rehearsal can lead scarcity of water and affect aquatic ecosystems in a damaging way.

Biodiversity Loss

Intensive monoculture practices and the use of high-yield crop forms can lead to a reduction in biodiversity, as they may not support diverse plant and animal species. This also responsible for the loss of wild cultivars.

Climate Change

Certain agricultural practices, such as rice cultivation and livestock farming, contribute to greenhouse gas emissions (methane and nitrous oxide), which contribute to climate change.

Soil Erosion

Unsustainable agricultural practices, like excessive plowing and deforestation, can result into soil erosion, which negatively impacts soil fertility and water quality, and can result in sedimentation of water bodies.

Resource Depletion

Over-extraction of water for irrigation and unsustainable farming practices can lead to the depletion of natural resources such as aquifers and fertile soils.

Sustainable Agriculture

Sustainable agricultural practices, such as organic farming (OF), agro-ecology and conservation agriculture, aim to minimize negative environmental impacts by promoting practices that are environmentally friendly, socially responsible and economically viable.

The relationship between agriculture and the environment is dynamical in nature and depends on the specific practices utilized. Sustainable and environmentally friendly agricultural practices are requisite to ascertain the long-term health of both agricultural systems and the broader environment. Balancing the need for food production with environmental conservation is an all-important dispute for global agricultural system.

Review of Literature

Various available literature about sustainable agriculture contemplates a raising consciousness regarding the demand to equilibrate food production along with environmental and social circumstances (Francis, 1990; Sonja, 2011; Youngberg & Harwood, 1989). Diverse research carries on to lead valuable perceptiveness related with sustainable agriculture systems.

Kareemulla et al. (2017) have analysed the agricultural sustainability across 19 major states in India for time-periods between 2001 and 2011. The agricultural sustainability index (ASI) was constructed by using the human development index methodology. The study found that ASI has increased over time for some states whereas the index decreased for others. The study concluded that appropriate government schemes for each dimension of agricultural sustainability may improve the performance of the concerned states.

Jatav and Naik (2023) measured agricultural sustainability studies at regional and district levels, implying environmental sustainability, social and economic security criteria. Here the authors coalesce the Pressure-State-Response framework and indicator theme to formulate composite ASI for various agro-climatic areas of India. The study reveals that the Trans Gengatic Plain Region ranks first position in agricultural sustainability and Eastern Himalayan Region ranks last. This type of study provides a better insight in connection with sustainability, its basic understanding, proper implementation and future design.

Suresh et al. (2022) considered 17 Indian states with specific indicator criteria for a time span of two decades. Here the workers applied normalized temporal data for preparation of indices for each of the selected indicators grouped under environmental, economic and social dimensions. The study witnessed that basic agricultural sustainability has amended in the Indian states. But the level of performance varies among states.

Gamage et al. (2023) described the various ill effects of conventional agriculture patterns on the environment. The ecological and environmental impingement can be downplayed by OF. OF practices encourage various recovery mechanisms. This minimizes environmental abasement and increases the quality of food. The authors also concluded that multiple OF technique and various innovative proficiencies (Biorefineries, Smart Nano fertilizers, Biochar, etc.) increase quality of agricultural products in a sustainable way.

Quantifying agricultural sustainability is very crucial for understanding the exact situation and for better implementation. Hyati et al. (2010) studied the troubles in this concern, factors influencing the measurement of sustainability and their interaction. They also pinpoint the basic touchstones for indicator selection and further implementation. They concentrated their study for national level farms and covey suitable recommendations.

Objectives of the Study

(i) To figure out the basics of agricultural sustainability.
(ii) To analyse the importance of agricultural sustainability.
(iii) Developing a composite index of agricultural sustainability through principal component analysis (PCA).
(iv) Conducting a comparative assessment of ASI among the states in India.
(v) To suggest suitable policy directives for improving the performance of the states.

Source of Data

All information in this research was exclusively obtained from secondary sources. Variables like forest coverage (FC), cropping intensity (CI), fertilizer use and OF were sourced from the Reserve Bank of India. Data on insured area (IA), agricultural income (AI), training opportunities and agricultural credit were collected from the Agricultural Statistics at a Glance 2022 report by the Ministry of Agriculture & Farmers Welfare, Government of India. Indicators like infant mortality rate (IMR) and sex ratio (SR) were taken from the report 'Women and Men in India' 2022. Education level data was acquired from the Global Data Lab.

Importance of Agricultural Sustainability

To support the ever-increasing demand of the growing population for food, we have to produce more and more nutrient sources. But the conventional agricultural practices result into various derangements, which lead to diminution of top soil, water shortage and subsequent mixing with pollutants, health hazards of the farmers, increase in the cost of farming practice and overall disbalance in the ecosystem.

Diverse path breaking initiatives has taken during the last few decades to furnish alternative scopes. Sustainable agriculture incorporates three principal facets namely overall health of the surrounding environment, economical gain and social impartiality (Alexander, 2017; Flora, 1992; Sonja, 2011; Youngberg & Harwood, 1989).

The non-conventional agricultural practices minimize the dependency on common natural resources like coal, petroleum, natural gas, etc. for the production of various chemical fertilizers. This measured use of natural resources also maintains a stable depository for the future generations. Sustainable agriculture provides a better economical position for marginal farmers and rural people with resources to a lesser extent. This practice also tackles natural calamities more effectively than the conventional one. Sustainable agriculture promotes social fairness and this allow for adequate admittance to affordable food sources for all. This exercise plays a pivotal role to save the natural wealth under various anthropogenic and natural fluctuations.

Dimensions of Sustainable Agriculture

Water is one of the primary natural resources that we have to consider for the establishment of life and agriculture and if mishandled, create an imbalance where sustainability is concerned. All the following measures should be adapted to use this source judiciously for the present and future generations:

(i) Ameliorate water conservation system properly.
(ii) Water storage scheme should be restored with recent management and planning.
(iii) Implementation of low volume irrigation style where possible.
(iv) Checking and maintenance of water quality from salinization and contaminations.
(v) A better weather forecasting system.

Present agricultural practice exclusively depends on petroleum. As this is a potent non-renewable energy source, its indiscriminate use creates a shortfall for future generations. Along with it increases the overall production costs, which subsequently creates an imbalance in socio-economic outcomes. In order to practice sustainable agriculture, we have to integrate various non-conventional energy sources like solar, air, hydroelectric power, etc. with the conventional one (Chel & Kaushik, 2011; Doval, 2018; Gorjian et al., 2022; Majeed et al., 2023).

The wholesale use of various chemical substances as fertilizer and allied substances makes the air quality degrading. Along with various other agricultural practices like dust particles from tillage, agricultural burning issues, mixing of pestilent substances with air and water sources produce huge problems. To extenuate these issues, we ought to follow these exercises:

(a) Maintenance of tillage.
(b) Incorporate techniques which minimizes agricultural runoff.
(c) Step-down top soil destruction.
(d) Practicing crop rotation, using cover crops, organic fertilizers, etc.
(e) Soil quality should be checked and maintained.
(f) Polyculture should be preferred than monoculture.
(g) Find out local varieties and their maintenance for better establishment within the local environmental considerations.

Circular Bio-Economy and Its Practices

The concept of circular bio-economy describes the means of converting and regulating our natural resources with industrial alignment through biological wealth to maintain sustainability. Fig. 11.1 describes it in a schematic presentation. The principle of reduce, recycle and reuse is considered to be the basic and appropriate way to establish circular bio-economy. Nutrient recovery plays a crucial role in this management (Adil et al., 2022; Nagarajan et al., 2021; Tišma et al., 2022).

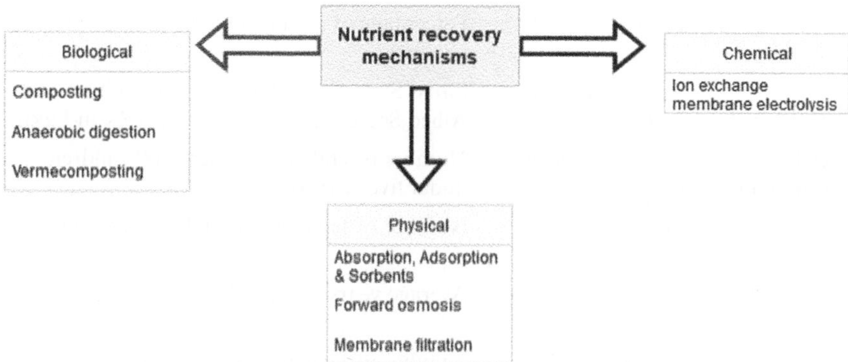

Fig. 11.1. Usual Techniques of Nutrient Recovery. *Source*: Compiled by the authors.

Construction of Agricultural Sustainability Index

To measure the agricultural sustainability, we constructed a composite index of agricultural sustainability index. We carefully selected a set of indicators based on a review of previous research studies on agricultural sustainability. A total of 12 indicators were chosen comprising four indicators from each dimension. These selections were based on the availability of data for the selected time period across the states in India. The details of the indicators are given in Panel I of Table 11.1 in category wise.

Table 11.1. Indicators for Measuring Agricultural Sustainability Index and Their Variations over the Years Across States.

Panel I: Indicators for Measuring Agricultural sustainability Index		
Dimension	**Indicator**	**Measure**
Environmental sustainability	*Forest Coverage (FC)*	Total geographical area declared as forest by the government (per square km).
	Cropping intensity (CI)	Percentage of gross cropped area to the net cultivated area.
	Chemical Fertilizer use intensity (CFI)	Used amount of nitrogen (N), phosphorous (P), and potassium (K) per hectare of cultivated area (kg/hectare).
	Organic Farming (OF)	Total cultivated and wild harvested area under organic certification (hectares).
Economic sustainability	*Food Grain Productivity (FP)*	Amount of Food grain production per hectare of sown area (kg/hectare)
	Agricultural Credit Facility (ACF)	Total amount of credit disbursed through Commercial banks, RRB's and Cooperative credit society (Rs. in lakh)
	Insured Area (IA)	Percentage of area insured under all insurance schemes.
	Agricultural Income (AI)	Gross State Value Added of Agriculture and Allied Sector at Constant prices (Rs. in Lakh).
Social Sustainability	*Infant mortality rate (IMR)*	Number of child deaths per 1000 children under five years of age.
	Sex ratio (SR)	Number of females per 1,000 males at birth.
	Level of Education (LE)	Average years of schooling.
	Training Facility (TF)	Number of farmers trained in farmers training under the extension reforms.

Forests are crucial for ecosystem services like climate regulation, water cycle, and carbon sequestration. They also provide livelihoods for marginalized communities through timber and non-timber products. Genetic diversity in forests is beneficial for crop breeding programs. Increasing forested areas can positively impact agricultural sustainability by enhancing crop diversity, equilibrium and resilience to natural disasters. Crop intensity reflects environmental aspects affecting agricultural sustainability, with recent decades showing increased global crop production due to more agricultural inputs. Excessive chemical fertilizer use can harm the environment, while OF offers an eco-friendly approach by promoting soil health and biodiversity preservation. OF supports sustainable agricultural practices.

Ensuring food grain production growth exceeds population growth is crucial for meeting rising food demand and maintaining food security. Improving food grain productivity (FP) enhances agricultural sustainability. Institutional credit is crucial for agricultural development, enabling farmers to invest in new technologies. In India, agricultural credit is vital for poverty reduction and has been emphasized since early planned development stages. Income generated from agriculture is a key indicator of sustainability, attracting more farmers to engage in agricultural activities and improve their lives.

The condition of a population's health is vital for agricultural sector development. IMR is a key social gauge affected by factors like agricultural progress, nutrition security, water access, and healthcare improvement. These elements directly affect infant well-being. Traditional farming practices have typically favoured men due to the physical demands, impacting gender ratios. Socioeconomic aspects like migration and labour in agriculture can influence this. Gender empowerment is pivotal for agricultural advancement.

The level of education (LE) is crucial for the success of any sector in the economy. Increasing education levels can improve agricultural productivity by enhancing knowledge, social capital and technical skills. Education is considered a positive factor in agricultural sustainability. Agricultural training is essential in boosting production, increasing income and securing the sustainability of the agricultural sector. Training programs can upgrade traditional farming practices to more efficient modern methods by teaching advanced techniques and technologies. This results in higher productivity and income for farmers. Sustainable farming practices and environmental stewardship promoted through agricultural training also contribute to the long-term viability of the agricultural sector, ensuring its continued role in food security and economic growth.

Methodology of Development of Agricultural Sustainability Index

The ASI consists of three dimensions like environmental, economic and social character (Jatav & Naik, 2023). The environmental dimension comprises four indicators like FC, CI, chemical fertilizer uses intensity and OF. On the other hand, economic dimension incorporates food grain productivity, agricultural credit, percentage of IA and income from agriculture and allied sector. Social dimension consists of IMR, SR, LE and agricultural training facility (TF).

In this study, to construct a composite index of agricultural sustainability, three-dimensional indices were formulated: environmental sustainability index (ESI), economic sustainability index (ECSI) and social sustainability index (SSI), utilizing the PCA method. The decision to divide the overall set of indicators into three dimension-indices served two main objectives. Firstly, it provided distinct meanings for each dimension index, thereby offering additional disaggregated information beneficial for policy formulation, as highlighted by Datta and Singh (2019). Secondly, for methodological clarity, given the high inter-correlation among indicators within each dimension, the estimation of dimension-indices was carried out separately before their combination to calculate the overall ASI.

In the initial stage, we formulated a dimension index for each dimension, namely the ESI, ECSI, and SSI, by incorporating various indicators specific to each dimension.

Before proceeding to construct the dimension index, the indicators within each dimension were 'normalized' to ensure better comparability of data. This normalization process was carried out using the UNDP goal-post method, as outlined below:

$$Z_i = \frac{(x_i - x_{\min})}{(x_{\max} - x_{\min})} \quad (1)$$

$$Y_i = \frac{(v_{\max} - v_i)}{(v_{\max} - v_{\min})} \quad (2)$$

Here equation (1) is used to indicators exhibiting a positive association. Here, Z_i represents the normalized value of the indicator for state $\boldsymbol{i}$, $\boldsymbol{x_i}$ denotes the corresponding pre-normalization figure, while $\boldsymbol{x_{max}}$ and $\boldsymbol{x_{min}}$ stand for the maximum and minimum values of the same indicator across all states, respectively.

Equation (2) is utilized for indicators such as chemical fertilizer use intensity (CFI) and IMR, which exert a negative influence on agricultural sustainability. Here, Y_i represents the normalized value of the indicator for the ith state, v_i denotes the corresponding pre-normalization figure and v_{max} and v_{min} denote the maximum and minimum values of the same indicator across all states, respectively. The normalized value of each indicator falls within the range of 0 and 1. A normalized indicator value of 0 signifies the lower end of the state's performance scale, while a value of 1 indicates the top end of the state's achievement for all individual categories of indicators.

The central challenge lies in assigning appropriate weights to the indicators, a crucial step to optimize the information extracted from a dataset included in an index. An effective composite index should integrate significant information from all indicators without exhibiting strong bias towards any particular indicator or set of indicators. This study applied PCA method as an indexing strategy to measure the degree of agricultural sustainability. By applying PCA method we constructed three-dimension index (ESI, ECSI and SSI) in the first stage. Whereas in the second stage the overall ASI was estimated as a mean value by considering these three-dimension index as indicators for overall ASI.

The ESI was formulated as a latent variable, incorporating indicators such as FC, CI, CFI, and OF. The dimension index was obtained through the weighted average form which is given as follows:

$$\text{ESI} = \frac{\sum_{j=1}^{4} \gamma_j^{\text{EV}} \cdot P_j^{\text{EV}}}{\sum_{j=1}^{4} \gamma_j^{\text{EV}}} \quad (3)$$

Here γ_j^{EV} represents the jth eigenvalue and P_j^{EV} denotes the jth principal component, respectively. It is important to note that the values of γ_j^{EV} gradually decline as the suffix increases. Since the first principal component (P_1^{EV}) explain the highest proportion of the variability in all explanatory variables, highest weightage (γ_1^{EV}) was assigned to it. The proportion of the variability represented by the principal component decreases as the suffix increases. Each principal component (P_j^{EV}) can be expressed as a linear combination of four indicators which is given as follows:

$$P_1^{EV} = \theta_{11}(FC) + \theta_{12}(CI) + \theta_{13}(CFI) + \theta_{14}(OF)$$
$$P_2^{EV} = \theta_{21}(FC) + \theta_{22}(CI) + \theta_{23}(CFI) + \theta_{24}(OF)$$
$$P_3^{EV} = \theta_{31}(FC) + \theta_{32}(CI) + \theta_{33}(CFI) + \theta_{34}(OF)$$
$$P_4^{EV} = \theta_{41}(FC) + \theta_{42}(CI) + \theta_{43}(CFI) + \theta_{44}(OF)$$

Here *P*s represent the principal components and θs denote loadings that are selected in such a way so that the principal components are uncorrelated.

By using FP, agricultural credit facility (ACF), IA and AI the ECSI was supposed to be determined. The weighted average form represented as

$$\text{ECSI} = \frac{\sum_{j=1}^{4} \gamma_j^{EC} \cdot P_j^{EC}}{\sum_{j=1}^{4} \gamma_j^{EC}} \quad (4)$$

where γ_j^{EC} (j = 1, 2, 3, 4) is the *j*th eigenvalue and P_j^{EC} is the *j*th principal component for economic sustainability.

The SSI was determined with the help of four indicators like IMR, SR, LE and TF. The weighted average form of the SSI can be represented as

$$\text{SSI} = \frac{\sum_{j=1}^{4} \gamma_j^{SO} \cdot P_j^{SO}}{\sum_{j=1}^{4} \gamma_j^{SO}} \quad (5)$$

Here γ_j^{SO} (j = 1, 2, 3, 4) denotes the *j*th eigenvalue and P_j^{SO} represents the *j*th principal component for social sustainability.

In the second stage we used arithmetic mean to construct the ASI with the help of the values of ESI, ECSI and SSI as constructed above and finally the ASI is represented as follows:

$$\text{ASI} = \frac{(\text{ESI} + \text{ECSI} + \text{SSI})}{3}$$

Results and Discussion

Status of Agricultural Sustainability Index across the States in India

This section delves into the analysis of the ASI across the states in India. A unique aspect of this study lies in the development of a composite ASI, which takes into account a range of environmental, economic and social indicators. Panel II of Table 11.1 presents the specific performance of agricultural sustainability for each

Table 11.1. Indicators for Measuring Agricultural Sustainability Index and Their Variations over the Years Across States.

Panel II: Year-Wise Variations of Agricultural Sustainability Index Across States				
States	**2016**	**2017**	**2018**	**2019**
Andhra Pradesh	0.361	0.424	0.423	0.492
Arunachal Pradesh	0.332	0.232	0.302	0.390
Assam	0.244	0.228	0.236	0.305
Bihar	0.232	0.250	0.209	0.281
Chhattisgarh	0.315	0.279	0.294	0.336
Goa	0.350	0.342	0.363	0.442
Gujarat	0.354	0.362	0.363	0.450
Haryana	0.325	0.411	0.400	0.488
Himachal Pradesh	0.408	0.363	0.378	0.446
Jammu & Kashmir	0.340	0.322	0.362	0.401
Jharkhand	0.277	0.253	0.265	0.324
Karnataka	0.362	0.346	0.353	0.450
Kerala	0.522	0.477	0.506	0.575
Madhya Pradesh	0.539	0.550	0.539	0.602
Maharashtra	0.493	0.482	0.473	0.532
Manipur	0.362	0.329	0.310	0.391
Meghalaya	0.255	0.222	0.233	0.310
Mizoram	0.326	0.299	0.343	0.441
Nagaland	0.364	0.315	0.339	0.436
Odisha	0.270	0.281	0.294	0.349
Punjab	0.471	0.476	0.456	0.543
Rajasthan	0.333	0.405	0.415	0.463
Sikkim	0.389	0.332	0.373	0.467
Tamil Nadu	0.458	0.573	0.595	0.609
Telangana	0.296	0.318	0.292	0.440
Tripura	0.363	0.300	0.310	0.405
Uttar Pradesh	0.404	0.423	0.420	0.464
Uttarakhand	0.251	0.292	0.307	0.404
West Bengal	0.464	0.424	0.442	0.480

Source: Compiled by the authors.

state in India. A higher ASI value indicates a relatively stronger performance by the state in terms of agricultural sustainability.

Panel II of Table 11.1 illustrates that, aside from 2016, Tamil Nadu attained the top position among the 29 states in terms of ASI. Conversely, the state Bihar secured last position across all years except 2017. Notably, agricultural sustainability improved for all states in 2019 compared to 2016. Among them, Haryana exhibited the most significant improvement, while West Bengal showed the least. A noteworthy observation is the continuous increase in ASI values for Gujarat, Odisha, Rajasthan, Tamil Nadu and Uttarakhand throughout the study period. Additionally, states like Madhya Pradesh, West Bengal, Kerala, Tamil Nadu, Uttar Pradesh, Punjab and Maharashtra consistently demonstrated strong performance over the entire duration. In contrast, Bihar, Assam, Jharkhand, Meghalaya and Odisha consistently ranked poorly in terms of ASI during the same period.

To analyse the performance of the states in a more concrete way, the values of ASI were divided into three categories. The value of the index greater than 0.5 indicates a higher level of agricultural sustainability. States with ASI values ranging from 0.3 to 0.5 are considered to have a medium level of performance in agricultural sustainability. States whose ASI values are lower than 0.3 are categorized as having a lower level of agricultural sustainability.

It was found that in the initial year of the study, 7 states namely Assam, Bihar, Jharkhand, Meghalaya, Odisha, Telangana and Uttarakhand out of 29 states experienced a lower level of agricultural sustainability. Twenty states (Chhattisgarh, Goa, Gujarat, Haryana, Himachal Pradesh, Jammu & Kashmir, Karnataka, Maharashtra, Manipur, Mizoram, Nagaland, Punjab, Rajasthan, Sikkim, Tamil Nadu, Tripura, Uttar Pradesh) were classified as having a medium level of agricultural sustainability. Kerala and Madhya Pradesh were the only states performing well regarding agricultural sustainability. In the year 2017, states like Arunachal Pradesh, Assam, Bihar, Chhattisgarh, Jharkhand, Meghalaya, Mizoram, Odisha, Tripura and Uttarakhand were categorized as low-performing states.

Among the 29 states, 17 states (Andhra Pradesh, Goa, Gujarat, Haryana, Himachal Pradesh, Jammu & Kashmir, Karnataka, Kerala, Maharashtra, Manipur, Nagaland, Punjab, Rajasthan, Sikkim, Telangana, Uttar Pradesh and West Bengal) are categorized as having a medium level of performance. Madhya Pradesh and Tamil Nadu are noted for their better performance in agricultural sustainability. In 2019, Bihar stands out as the only low-performing state. On the other hand, states like Kerala, Madhya Pradesh, Maharashtra, Punjab and Tamil Nadu belong to the group showing a higher level of agricultural sustainability. The remaining states are categorized as having a medium level of performance. When comparing the performance categories of states, it was observed that Assam, Jharkhand, Meghalaya, Odisha, Telangana and Uttarakhand transitioned from being low performers in 2016 to the medium level of performance in 2019.

Conclusion

The research focused on the importance of agricultural sustainability and formulated a composite index to evaluate the performance of states in India from 2016 to 2019. It revealed significant disparities in agricultural sustainability among the states. It is important to mention that agricultural sustainability improved for all states in 2019 compared to 2016. Progress in agricultural sustainability increased for all states during this period, with Haryana making the most substantial gains and West Bengal showing the least improvement. It has been found for developing countries like India that we have to accomplish a long run towards sustainability. To encourage the sustainable agricultural practice, the National Policy on Agriculture and budgetary allocation should be modified so that we can implement these theoretical practices much more to reality.

References

Adil, M., Nasir, A., Sikandar, A., & Khan, N. M. (2022). *Recent advances in circular bioeconomy* (pp. 59–84). Springer E-Books.

Alexander, W. (2017). *Agroecological practices for sustainable agriculture: Principles, applications, and making the transition*. World Scientific.

Chel, A., & Kaushik, G. (2011). Renewable energy for sustainable agriculture. *Agronomy for Sustainable Development*, *31*(1), 91–118. https://doi.org/10.1051/agro/2010029

Close & Australian Government, Department of The Environment, Water, Heritage, And the Arts. (2010). Ecosystem services: Key concepts and applications. Canberra, A.C.T.: Australian Govt., Dept. of the Environment, Water, Heritage, and the Arts.

Costanza, R., d'Arge, R., de Groot, R., Farber, S., Grasso, M., Hannon, B., Limburg, K., Naeem, S., O'Neill, R. V., Paruelo, J., Raskin, R. G., Sutton, P., & van den Belt, M. (1997). The value of the world's ecosystem services and natural capital. *Nature*, *387*(6630), 253–260. https://doi.org/10.1038/387253a0

Datta, S. K., & Singh, K. (2019). Variation and determinants of financial inclusion and their association with human development: A cross-country analysis. *IIMB Management Review*, *31*(4), 336–349.

Doval, C. (2018). What is sustainable agriculture? Retrieved from Sustainable Agriculture Research & Education Program website: https://sarep.ucdavis.edu/sustainable-ag

Flora, C. B. (1992). Building sustainable agriculture. *Journal of Sustainable Agriculture*, *2*(3), 37–49. https://doi.org/10.1300/j064v02n03_04

Francis, C. A. (1990). Sustainable agriculture. *Journal of Sustainable Agriculture*, *1*(1), 97–106. https://doi.org/10.1300/j064v01n01_08

Gamage, A., Gangahagedara, R., Gamage, J., Jayasinghe, N., Kodikara, N., Suraweera, P., & Merah, O. (2023). Role of organic farming for achieving sustainability in agriculture. *Farming System*, *1*(1), 100005.

Gorjian, S., Fakhraei, O., Gorjian, A., Sharafkhani, A., & Aziznejad, A. (2022). Sustainable food and agriculture: Employment of renewable energy technologies. *Current Robotics Reports*, 3(3), 153–163. https://doi.org/10.1007/s43154-022-00080-x

Hayati, D., Ranjbar, Z., & Karami, E. (2011). Measuring agricultural sustainability. Biodiversity, *Biofuels*, *Agroforestry* and *Conservation Agriculture*, 73–100.

Jatav, S. S., & Naik, K. (2023). Measuring the agricultural sustainability of India: An application of pressure-state-response model. *Regional Sustainability*, *4*(3), 218–234.

Kareemulla, K., Venkattakumar, R., & Samuel, M. P. (2017). An analysis on agricultural sustainability in India. *Current Science*, *112*(2), 258–266.

Kumara, T. M. K., Pal, S., Chand, P., & Kandpal, A. (2023). Carbon sequestration potential of sustainable agricultural practices to mitigate climate change in Indian agriculture: A meta-analysis. *Sustainable Production and Consumption*, *35*, 697–708. https://doi.org/10.1016/j.spc.2022.12.015

MacDonald, G. K., Bennett, E. M., Potter, P. A., & Ramankutty, N. (2011). Agronomic phosphorus imbalances across the world's croplands. *Proceedings of the National Academy of Sciences*, *108*(7), 3086–3091.

Majeed, Y., Khan, M. U., Waseem, M., Zahid, U., Mahmood, F., Majeed, F., & Raza, A. (2023). Renewable energy as an alternative source for energy management in agriculture. *Energy Reports*, *10*, 344–359.

Matson, P. A., Parton, W. J., Power, A. G., & Swift, M. J. (1997). Agricultural intensification and ecosystem properties. *Science*, *277*(5325), 504–509.

Nagarajan, D., Lee, D. H., & Chang, J. (2021). *Circular bioeconomy: An introduction* (pp. 3–23). Elsevier EBooks. https://doi.org/10.1016/b978-0-12-821878-5.00006-4

Sanaullah, M., Afzal, T., Shahzad, T., & Wakeel, A. (2019). *Carbon sequestration for sustainable agriculture* (pp. 469–500). Springer EBooks. https://doi.org/10.1007/978-3-030-23169-9_15

Singh, S., & Nayak, S. (2020). Development of sustainable livelihood security index for different agro-climatic zones of Uttar Pradesh, India. *Journal of Rural Development*, *39*(1), 110–129.

Smith, A., & Serna, I. (2023). The importance of sustainable agriculture. Retrieved from Plug and Play Tech Center.

Sonja, B. (2011). Sustainable agriculture learn science at Scitable. Retrieved from Nature.com.

Suresh, A., Krishnan, P., Jha, G. K., & Reddy, A. A. (2022). Agricultural sustainability and its trends in India: A macro-level index-based empirical evaluation. *Sustainability*, *14*(5), 2540.

Tišma, S., Farkaš, A., Pisarović, A., Funduk, M., & Tolić, I. (2022). *Economics, ecology, and policy for the bioeconomy*. Routledge EBooks. https://doi.org/10.4324/9781003223733

Youngberg, G., & Harwood, R. (1989). Sustainable farming systems: Needs and opportunities. *American Journal of Alternative Agriculture*, *4*(3–4), 100–100.

Chapter 12

Assessing the Relationship Between Crop Production, Livestock Production, and CO_2 Emissions: A Panel Cointegration Analysis

Ujjal Protim Dutta and Aliul Islam

Department of Humanities and Social Sciences, National Institute of Technology Silchar, Assam, India

Abstract

During the late 1900s, there was a notable trend of rapid urbanization worldwide, coinciding with a rise in atmospheric carbon dioxide levels. Agricultural production driven by expanded trade is recognized as a major contributor to global pollution and biodiversity depletion. Additionally, tropical deforestation resulting from agricultural activities significantly impacts global greenhouse gas emissions. This chapter aims to examine the interrelationships between crop production, livestock production, transport services, and CO_2 emissions from 1998 to 2019. To achieve this goal, the study begins by conducting stationary tests to determine the order of integration for the variables under consideration. Following this, panel unit root tests are employed, and subsequently, panel cointegration tests are conducted to identify any long-term relationships among the selected variables. The findings reveal a significant long-term relationship among crop production, livestock production, transport services, and CO_2 emissions.

Keywords: Crop production; livestock production; CO_2 emission; agricultural development; panel unit root; panel cointegration

Sustainable Agricultural Practices: Economic and Environmental Implications, 177–190

doi:10.1108/978-1-83608-336-820241012

Introduction

Rising levels of carbon dioxide (CO_2) pose escalating environmental challenges. This uptick in CO_2 emissions is often observed alongside irregular and rapid industrialization, urbanization, imbalanced economic growth, population expansion, and heightened energy consumption (Moutinho et al., 2018). Particularly concerning population growth and industrialization, the inevitable escalation of agricultural production is necessary to ensure food security and maintain a consistent supply of raw materials for industries. However, this surge in production amplifies CO_2 emissions, with the actual increase often surpassing initial expectations due to improper practices (Erbas & Solakoglu, 2017; Schneider & Smith, 2009). Like other industries, agriculture depends on energy as a fundamental input for its operations. Energy constitutes a vital component in agricultural operations, serving as a necessary input across various activities. From plowing fields to preparing fodder, irrigating crops, harvesting, drying, and grinding grains, energy consumption is integral to agricultural processes. Within the agricultural sector, non-renewable energy sources like fossil fuels, electricity, natural gas, and coke play pivotal roles. The power machinery and equipment, provide heating or cooling for farm buildings, and illuminate agricultural spaces. Furthermore, non-renewable energy sources are indirectly utilized for the production of fertilizers, machinery, and chemicals essential for agricultural processes both on and off the farm.

Livestock production, particularly in agriculture, stands out as a significant contributor to global warming, as highlighted by researchers (Malhi et al., 2021; Pata, 2021). At the same time, both crop cultivation and livestock rearing emerge as the sectors most vulnerable to the adverse effects of climate change (Mendelsohn, 2009). The increasingly prevalent extreme arid conditions, a consequence of climate change, lead to reduced crop yields by impeding nutrient mobility and fostering soil salinity, rendering it infertile and ultimately barren (Arora, 2019). The decrease in crop production becomes more pronounced with the escalation of CO_2 emissions, particularly in the absence of carbon fertilization (Ackerman & Munitz, 2016). However, agriculture serves as the cornerstone of meeting society's essential food needs, albeit it must be pursued without compromising global food security or ecological balance (Bennetzen et al., 2016). Amidst unprecedented climate change, both crop and livestock production are expected to face declining yields (Malhi et al., 2021). The pressure on agriculture arises from human activities that undermine environmental quality and diminish the available arable land for livestock farming (Rojas-Downing et al., 2017).

Undoubtedly, misguided agricultural practices, including cultivating in unsuitable areas to boost production, reliance on pesticides and chemical fertilizers, inefficient irrigation methods, improper soil management, errors in the use of plant hormones, stubble burning, and improper disposal of animal waste into the soil, contribute to increased CO_2 emissions from crop production (Waheed et al., 2018). Similarly, improper grazing practices, ineffective fertilization techniques, destruction of pastures, and inaccuracies in feeding methods in animal husbandry can also contribute to increased CO_2 emissions. Moreover, such practices raise concerns about environmental degradation and the long-term sustainability of agriculture, as well as food security. While industrial activities are the primary contributors to CO_2 emissions, the spatial distribution and trends in agricultural CO_2 emissions are also significant factors in climate change.

The IPCC's 2023 report highlights that in 2019, approximately 22% of greenhouse gas emissions responsible for global warming originated from agriculture, forestry, and other land use activities. Unlike ancient times when agriculture relied less on fossil fuels and had a smaller environmental footprint due to lower population levels, today's industrial farming necessitates extensive energy usage for tasks such as mechanized plowing over large land areas. The production and operation of agricultural machinery, driven by high demand, further contribute to fossil fuel consumption in the automobile industry. Moreover, the widespread use of chemical fertilizers in modern agriculture, aimed at increasing output from limited land resources, entails the production of these compounds using fossil energy. This not only degrades the atmosphere but also harms soil organisms and organic matter. Additionally, agriculture's reliance on fossil energy extends to irrigation, refrigerated storage of produce, and land clearing for expanded agricultural activities, leading to accelerated deforestation.

The urgent need to address global warming and reduce CO_2 emissions has garnered significant attention worldwide. Governments and policymakers are calling for empirical research and practical measures to achieve environmental sustainability, which has become a top priority for environmentalists, researchers, policymakers, and governments alike. Despite concerted efforts, achieving targeted environmental goals is proving challenging due to the persistent growth of the economy, which often comes at the expense of environmental degradation through activities such as agriculture, forestry, and manufacturing. Previous research on environmental quality has identified several key variables that influence environmental conditions, including urbanization, globalization, tourism, industrialization, direct expenditures, literacy, public spending, and inclusive finance (Usman et al., 2021). However, there exists a discernible void in the existing literature concerning the correlation between agricultural activities and CO_2 emissions.

In light of this context, this study aims to explore the association between crop production, livestock production, transportation services, and CO_2 emissions across ten selected Asian countries from 1998 to 2019. The remainder of the work is structured as follows: following the introduction, the second section delves into an examination of the existing literature. The third section presents an analysis of the global and regional trends in emissions from crop and livestock production. In the fourth section, we detail the methodology employed, while the fifth section comprises the results and discussion. Finally, the sixth section provides the conclusion of the study.

Literature Review

The Climate Watch Report (2023) emphasizes that climate change is primarily driven by human-induced emissions, with CO_2 being a predominant factor. In recent decades, the surge in economic activities has been accompanied by a notable increase in greenhouse gas emissions, exacerbating the frequency and intensity of natural disasters on a global scale. This assertion is supported by studies conducted by Apergis and Ozturk (2015), Škare et al. (2020), and Ullah et al. (2021), which underscore the correlation between economic expansion and heightened emissions. As economic activities continue to expand, the consequential rise in greenhouse gas emissions poses significant challenges, exacerbating the impact of climate change and leading to widespread natural calamities worldwide.

Over the last 20 years, numerous empirical studies in the literature have extensively examined various factors contributing to the dynamics of CO_2 emissions. These factors encompass economic growth, energy utilization (both renewable and non-renewable), trade patterns, transportation systems, urban development, financial advancements, foreign direct investment, and tourism activities. Furthermore, despite the significance of all the factors mentioned in influencing CO_2 emissions, research on the impact of agricultural activities has been disproportionately minimal within academic, economic, and regulatory discourse.

The World Bank (2013) highlights that agriculture contributes significantly to carbon emissions and it ranges from one-fourth to one-third of total emissions. Within this sector, on-farm activities account for 10–12% of global emissions, while land-use changes to cropland contribute 12–20% (World Bank, 2013). It is imperative to adopt low-carbon agricultural practices to safeguard the environment and foster economic development (Fan et al., 2015; Rehman et al., 2019; Xiao et al., 2021).

In the literature, some empirical studies have examined the relationship between the agricultural sector and CO_2 emissions. However, these investigations have not yielded a definitive conclusion. The variation in results across studies can be attributed to differences in datasets, selected countries or country groups, time periods, and methodological approaches. Studies exploring the connection between CO_2 emissions and the agricultural sector have produced two distinct conclusions.

One perspective suggests a linear relationship between the agricultural sector and CO_2 emissions. For instance, Liu et al. (2017) employed fully modified ordinary least squares-dynamic ordinary least squares (FMOLS-DOLS) and vector error correction model (VECM) methodologies to analyze the interplay among renewable energy, agriculture, and the environment in BRICS countries. Their findings indicate a positive impact of agriculture on CO_2 emissions, establishing a unidirectional causality from agriculture to CO_2 emissions over the long term. Similarly, Zhangwei and Xungang (2011) identified a robust relationship between agricultural CO_2 emissions and agricultural economic growth. Their findings indicate that as agricultural growth intensifies, both agricultural CO_2 emissions and total CO_2 emissions increase. Likewise, Sarkodie and Owusu (2016) investigated the link between CO_2 emissions and the agricultural sector in Ghana, utilizing autoregressive distributed lag (ARDL) and VECM models. Their analysis confirmed the presence of a causal relationship between agriculture and CO_2 emissions. By examining variables such as annual changes in agricultural areas, livestock space allocation, and selected product production, the study revealed that all these factors contribute to an increase in CO_2 emissions. Similarly, Balsalobre-Lorente et al. (2019) applied FMOLS, DOLS, and panel causality tests specifically to BRICS countries. They contended that agricultural activities exert pressure on the environment, concluding that there exists a unidirectional causality from agriculture to CO_2 emissions. In line with the above-mentioned literature, Ullah et al. (2018) employed ARDL, Johansen cointegration test, and Granger causality test in their research. Their investigation revealed that various agricultural activities in Pakistan are associated with heightened CO_2 emissions. Moreover, their study uncovered a bidirectional causality between agriculture and CO_2 emissions. In a study conducted by Pata (2021) focusing on BRICS countries from 1971 to 2016, it was affirmed that there exists a two-way causality between agriculture and environmental pollution. The research highlights the dual role of agriculture as both a victim of and a contributor to environmental pollution. Agricultural practices contribute to carbon emissions through various means, including the burning of farmlands, drainage of wetlands, utilization of agrochemicals, rearing of ruminant animals producing methane, and reliance on fossil energies throughout the production process.

The second perspective suggests that increased agricultural production either reduces CO_2 emissions or has little impact on emissions. For instance, Ben Jebli and Ben Youssef (2017) investigated the role of agriculture and renewable energy in mitigating CO_2 emissions using the FMOLS-DOLS method over the long term, and they assessed causality using the VECM Granger causality method. Their study, focusing on North African countries from 1980 to 2011, concluded that an increase in agricultural income leads to a reduction in CO_2 emissions in the long term. Additionally, they identified a bilateral causality relationship between agricultural income and CO_2 emissions in both the short and long terms. Dogan (2016) investigated the relationship between agriculture and CO_2 emissions in Turkey utilizing the ARDL model. Their findings indicate that an increase in both short- and long-term crop production leads to a reduction in CO_2 emissions. Similarly, Dufour et al. (2009) employed three greenhouse gases (GHGs) as proxies for environmental degradation and utilized the ARDL model. Their analysis revealed a negative correlation between climate change and agricultural productivity.

Given the inconclusive nature of research on the relationship between agricultural activities and CO_2 emissions, it becomes crucial to thoroughly investigate this association. Despite numerous studies in this field, there remains a lack of definitive conclusions regarding the extent to which agricultural practices contribute to CO_2 emissions. Therefore, it becomes imperative to delve deeper into this area to gain a clearer understanding of the complex interplay between agricultural activities and their environmental impact. Moreover, while previous research has explored the correlation between various factors and environmental quality, there exists a noticeable gap in our understanding of the specific contributions of different agricultural components to CO_2 emissions. While it's known that agriculture as a whole has a significant impact on CO_2 emissions, the specific mechanisms and factors within agricultural practices that drive these emissions remain less understood. By filling this gap in knowledge, researchers can provide valuable insights that can inform more targeted and effective mitigation strategies to reduce CO_2 emissions from agricultural activities. Therefore, further research focusing on dissecting the specific contributions of various agricultural components to CO_2 emissions is essential for informing sustainable agricultural practices and environmental policy-making.

An Overview of Global and Regional Trends of Green House Gas Emission from Crop and Livestock Production

The agricultural sector, like any other industry, relies on energy for its operations. However, it also generates significant non-CO_2 emissions, including methane and nitrous oxide, which are potent GHGs. In 2018, these emissions amounted to 5.3 gigatons of CO_2 equivalent, according to the FAO (2020). Livestock production accounted for approximately two-thirds of this total, as illustrated in Fig. 12.1. These emission shares have shown a consistent decline in both crop production and livestock production. In 1998, the emission shares were 2.15% and 9.45% for crop and livestock production, respectively. Over time, these shares have gradually decreased, with the figures dropping to 1.66% and 7.64% for crop and livestock production, respectively, by 2019. In addition to this, Fig. 12.1 also illustrates the regional trends in CO_2eq emissions from agricultural activities, specifically crop and livestock production, across different continents. The data, sourced from FAOSTAT, covers the years 1998, 2009, and 2019 from left to right, presented as emissions share (CO_2eq) (AR5). In Africa, there is a noticeable upward trend in CO_2eq emissions, with a 0.29% increase in crop production emissions and a substantial 3.02% increase in livestock production emissions from 1998 to 2019. Northern America also exhibits an upward trend in CO_2eq emissions, with a slight 0.03% increase in crop production emissions and a modest 0.14% increase in livestock production emissions from 1998 to 2019. Similarly, South America shows an upward trend in CO_2eq emissions, with a notable 0.51% increase in crop production emissions and a substantial 5.96% increase in livestock production emissions from 1998 to 2019.

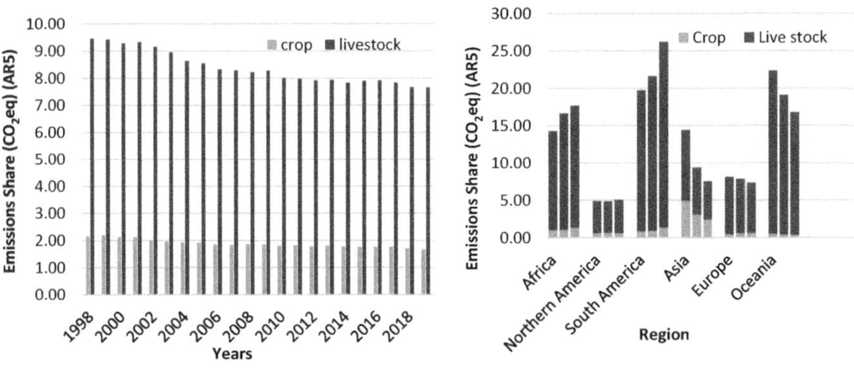

Fig. 12.1. Yearly Emission from Crop and Livestock Components of Agriculture, 1998–2019 and Trends of Regional Emission from Crops and Livestock Production in the Years 1998, 2009 and 2019. *Source*: Drawn by the authors using the database of FAOSTAT (2020).

Materials and Method

Specification of the Model

To examine the association between crop production, livestock production, transport services and CO_2 emission, this chapter relied on the following equation:

$$CO_{2it} = \beta_1 \text{Crop Production}_{i,t} + \beta_2 \text{Livestock Production}_{i,t} + \beta_3 \text{Transport Sevices}_{i,t} + \vartheta_i + \varepsilon_{i,t} \quad (1)$$

where crop production, livestock production and transport services are explanatory variable of the study. CO_2 emission has been considered as the dependent variable of the study and $\varepsilon i, t$ is the disturbance term.

Data

To achieve the study's objective, a panel comprising India, China, Indonesia, Pakistan, Bangladesh, Japan, Philippines, Turkey, Thailand, and Myanmar spanning from 1998 to 2019 was utilized. In 2019, when evaluating total CO_2 emissions (excluding agricultural sources), China, India, and Japan emerged as the primary contributors in the Asian region. However, when specifically examining emissions from livestock production, Indonesia, Turkey, and Pakistan were identified as the top emitters. Similarly, concerning emissions from crop production, India, Turkey, and Bangladesh ranked highest. Moreover, in terms of emissions attributable to commercial service imports, Bangladesh, Myanmar, and Pakistan stood out as the leading emitter nations.

The lack of consistent data for the variables of interest before 1998 posed a constraint, leading to the adoption of data from this specific period to ensure uniformity across the variables. Annual data for crop production, livestock production, transport services, and CO_2 emissions were sourced from the World Bank's WDI database. The study anticipates identifying statistically significant relationships among the chosen variables, thereby contributing to a comprehensive understanding of their interconnections.

Variable Descriptions

CO_2 emissions (kt): CO_2 emission has been considered as dependent variable of the study. CO_2 emissions data encompass gases generated from fossil fuel combustion and cement production, with measurements typically expressed in kilotons (kt). It's important to note that these figures do not account for emissions from land use activities like deforestation. Additionally, CO_2 emissions are commonly quantified and reported in terms of elemental carbon. Data on CO_2 emissions have been extracted from the World Development Indicator database.

Crop production index: The crop production index reflects agricultural output for each year compared to the base period of 2014–2016. This index encompasses all crops except fodder crops. Regional and income group aggregates for the FAO's production indexes are derived from the original values measured in international dollars and standardized to the base period of 2014–2016. Data on crop production index has been obtained from World Development Indicator Database.

Livestock production index: The livestock production index comprises meat and milk from various sources, including dairy products like cheese, along with eggs, honey, raw silk, wool, and hides and skins. Like crop production index and CO_2 emission, data on livestock production index has been taken from World Development Indicator Database.

Transport services: The indicator "Transport Services (% of Commercial Service Imports)" encompasses all transportation services – such as sea, air, land, internal waterway, space, and pipeline – provided by residents of one economy to those of another. These services include the transportation of passengers and goods (freight), as well as the rental of carriers with crew, along with associated support and auxiliary services. Data on transport services has been taken from World Bank Development Indicator Database.

Methods

Stationary Test

To determine the stationarity of the given data, two tests were conducted: the Levin, Lin, and Chu (LLC) test and the Im, Pesaran, and Shin (IPS) test. Both tests are extensions of the augmented Dickey–Fuller (ADF) test tailored for time series data, addressing the complexities of the limiting distribution in individual time series techniques. The LLC test, proposed by LLC (2002), assumes the presence of a unit root in the null hypothesis, with the alternative hypothesis suggesting its absence. It assumes a common unit root process across all panel members. In contrast, the IPS test assumes individual unit root processes among panel members. The null hypothesis for the IPS test is that the coefficient is $H_0{:}pi = 0$ for all *i*'s (for all panel members), while the alternative hypothesis suggests $H_1{:}pi < 0$ for a few *i*'s for at least one *i*. The structure of both LLC and IPS tests can be expressed through the following ADF regression equation:

$$\Delta y_{i,t} = \alpha_i + \rho y_{i,t-1} + \sum\nolimits_{j=1}^{p_i} \alpha_j \Delta y_{i,t-j} + \varepsilon_{i,t} \tag{2}$$

where p_i = number of lags and $\varepsilon_{i,t}$ shows disturbance terms.

Cointegration Analysis

Following the investigation of data stationarity, the subsequent step involves exploring the potential cointegrating relationship among the selected variables. To analyze this relationship, Pedroni's (1999) cointegration tests are employed, which consist of two types: panel tests and group tests. The panel tests, also known as "within dimension," encompass four test statistics: Panel-v, Panel-rho, Panel PP-statistic, and Panel ADF-statistic. These statistics are derived from estimated residuals obtained from a regression model where the first differences of the dependent variable and explanatory variables are regressed on individual-specific intercepts, trend terms, lagged variables, and error terms. On the other hand, the group tests, referred to as "between dimension," include three test statistics: Group-rho, Group PP-statistics, and Group ADF-statistics. Similar to panel tests, these statistics are derived from estimated residuals but capture relationships across panel groups rather than within them. Overall, these test statistics provide insights into the presence of a cointegrating relationship among the variables, shedding light on potential long-term equilibrium relationships.

The seven test statistics are based on estimated residuals from the following regression specification:

$$y_{i,t} = \alpha_i + \delta_i t + \beta_1 X_{1,i,t} + \beta_2 X_{2,i,t} + \cdots + \beta_n X_{n,i,t} + \varepsilon_{i,t} \tag{3}$$

where αi = fixed effects specific to countries and δi indicates a deterministic time trend. The term $\varepsilon_{i,t}$ shows the estimated residuals and the given regression equation is implemented on it to test the null hypothesis of no cointegration:

$$\varepsilon_{i,t} = \rho_i \varepsilon_{i,t-1} + \mu_{i,t} \tag{4}$$

On the one hand, the "Within Dimension" approach assumes the null hypothesis, i.e., H_0:$\rho i = 1$ for all i against the alternative hypothesis, i.e., H_1:$\rho i < 1$ for all i while on the other "Between Dimension" approach assumes null hypothesis, i.e., H_0:$\rho i = 1$ for all i against the alternative hypothesis, i.e., H_1:$\rho i < 1$ for at least one i.

Results and Discussion

Results of Stationarity Test

In the process of selecting an appropriate estimation method for the study, a critical step involves evaluating the stationarity characteristics of the variables under consideration. As detailed in Section 4.3.1 of the study, this evaluation was conducted through the application of two distinct tests: the LLC test and the IPS test. These tests are widely recognized in econometric analysis for their effectiveness in assessing the stationarity of time series data. The results of these tests, presented in Table 12.1, offer valuable insights into the behavior of the variables.

Both the LLC and IPS tests were applied to all pertinent variables in their original levels, as well as in their first differences. This comprehensive approach allowed for a thorough examination of the stationarity properties of each variable across different transformations. Upon analysis, it was observed that the variables exhibited non-stationarity when examined in their original levels. However, upon taking the first differences of the variables, they displayed a stationary behavior.

This finding holds significant implications for the subsequent steps of the study. By establishing the stationarity of the variables in their first differences, it provides a solid foundation for the choice of appropriate estimation methods in

Table 12.1. Outcomes of Unit Root and Cointegration Tests.

Results of LLC and IPS Unit Root Tests

	LLC		IPS	
	LD	**FD**	**LD**	**FD**
Lnco2	0.8201	−6.4237*	0.9888	−6.1159*
Lncp	−1.4754	−6.4936*	−0.5305	−7.2579*
Lnlp	2.6906	−11.5963*	4.4337	−9.5145*
Lntm	−1.3355	−8.0410*	−0.9306	−9.1092*

Results of Pedroni Cointegration Tests

	Within Dimension		Between Dimension	
	Panel	***P*-value**	**Group**	***P*-Value**
v-statistic	4.1611	0.0000*	–	–
P-statistic	0.9468	0.8281	1.8113	0.9650
t-statistic (non-parametric)	−2.1031	0.0177*	−2.9432	0.0016*
t-statistic (parametric)	−1.8288	0.0337**	−3.3311	0.0004*

Source: Authors' own calculation using E-views Software.
Note: LD: level data; FD: first differenced data; LLC: Levine–Lin–Chu statistics; IPS: Im–Pesaran–Shin statistics.
Note: ***, ** and * represent 10%, 5% and 1% level of significance.

subsequent analyses. Stationarity is a fundamental assumption in many econometric models, and the confirmation of stationarity in the first differences of the variables ensures the validity of these models. Furthermore, the recognition of non-stationarity in the original levels of the variables highlights the importance of appropriate data pre-processing techniques. Non-stationary data can introduce biases and inaccuracies in econometric analyses, potentially leading to erroneous conclusions. Therefore, the identification of non-stationarity underscores the necessity of employing techniques such as first differencing to ensure the reliability of the subsequent analyses.

In summary, the application of the LLC and IPS tests to evaluate the stationarity characteristics of the variables in the study represents a crucial step in the analytical process. The confirmation of stationarity in the first differences of the variables provides a robust foundation for the selection of appropriate estimation methods and ensures the reliability of the subsequent econometric analyses.

Results of Cointegration Test

The outcomes of Pedroni's cointegration test are presented in Table 12.1, revealing significant findings regarding the presence of cointegrating relationships among the variables under examination. Within the "Within Dimension Approach," it was observed that two-test statistics rejected the null hypothesis of no cointegrating relationship at a 1% significance level, while one-test statistic did so at a 5% significance level. Similarly, within the "Between Dimension Approach," two-test statistics rejected the null hypothesis of no cointegrating relationship at a 1% significance level. Overall, the results indicate that four out of the seven test statistics rejected the null hypothesis of no cointegration among the selected variables at a 1% significance level. These findings underscore the presence of significant cointegrating relationships among the variables, providing valuable insights into their long-term equilibrium dynamics.

Conclusion

Environmental challenges such as global warming, water pollution and deforestation pose grave threats to the sustainable utilization of natural resources, endangering the well-being of present and future generations. Despite concerted efforts through various international meetings and agreements aimed at mitigating these issues and achieving Sustainable Development Goals (SDGs), progress has been insufficient. Despite commitments made in landmark agreements like the Kyoto Protocol and Paris Conference to reduce emissions, the trajectory of CO_2 emissions continues to rise, particularly in the developing world. This persistent rise in CO_2 emissions despite international commitments highlights the urgency for new research endeavors. There is a critical need to delve deeper into understanding the complex factors driving environmental degradation. Additionally, further research is required to identify gaps in current policies and to formulate more effective strategies to address environmental challenges. In response to this imperative, the current study embarks on an investigation into the impact of agricultural activities on CO_2 emissions in selected Asian countries.

The agricultural sector plays a vital role in both food production and employment generation, particularly in developing economies. As populations grow, there is a significant increase in agricultural productivity to meet the rising demand. However, these intensified agricultural activities can have adverse effects on the environment, disrupting the balance by emitting CO_2 and other GHGs. This chapter aims to examine the interrelationships between crop production, livestock production, transport services, and CO_2 emissions from 1998 to 2019. To achieve this goal, the study begins by conducting stationary tests to determine the order of integration for the variables under consideration. Following this, panel unit root tests are employed, and subsequently, panel cointegration tests are conducted to identify any long-term relationships among the selected variables. The findings reveal a significant long-term relationship among crop production, livestock production, transport services, and CO_2 emissions.

Efforts to mitigate the impact of agriculture on CO_2 emissions must prioritize enhancing energy efficiency. This entails transitioning from fossil fuel usage to renewable energy sources and accelerating the adoption of eco-friendly technologies. Increasing the adoption of renewable energy sources and technologies within agricultural practices is crucial for mitigating the impact of CO_2 emissions on the environment. This includes integrating solar-powered solutions such as solar dryers, solar pumps, and irrigation systems, as well as harnessing wind power and other renewable options. By embracing these sustainable practices, the agricultural sector can significantly reduce its carbon footprint while enhancing productivity.

Moreover, it's essential for selected countries included in this study to prioritize the integration of climate change considerations into regional sectoral plans. This entails developing strategies and policies that address the specific challenges posed by climate change within agriculture, ensuring resilience and sustainability in the face of evolving environmental conditions. Additionally, there is a pressing need for increased engagement by civil society organizations to raise awareness among the public about the impacts of CO_2 emissions on agricultural production. These organizations play a vital role in advocating for sustainable practices and policies, mobilizing communities, and fostering dialogue with governments to promote responsive action. By fostering a collaborative approach between stakeholders, we can work toward building a more sustainable agricultural sector that effectively addresses the challenges of climate change while ensuring food security and environmental conservation.

In addition, given the significant contribution of certain countries to agricultural emissions, it's imperative for these nations to enact or enhance policies aimed at optimizing pesticide and chemical fertilizer usage. Moreover, preventing the conversion of unsuitable forest land for agricultural purposes should be a key focus of these policies.

References

Ackerman, F., & Munitz, C. (2016). A critique of climate damage modeling: Carbon fertilization, adaptation, and the limits of FUND. *Energy Research & Social Science, 12*, 62–67.

Apergis, N., & Ozturk, I. (2015). Testing environmental Kuznets curve hypothesis in Asian countries. *Ecological Indicators, 52*, 16–22.

Arora, N. K. (2019). Impact of climate change on agriculture production and its sustainable solutions. *Environmental Sustainability, 2*(2), 95–96.

Balsalobre-Lorente, D., Driha, O. M., Bekun, F. V., & Osundina, O. A. (2019). Do agricultural activities induce carbon emissions? The BRICS experience. *Environmental Science and Pollution Research, 26*, 25218–25234.

Ben Jebli, M., & Ben Youssef, S. (2017). Renewable energy consumption and agriculture: Evidence for cointegration and Granger causality for Tunisian economy. *International Journal of Sustainable Development & World Ecology, 24*(2), 149–158.

Bennetzen, E. H., Smith, P., & Porter, J. R. (2016). Decoupling of greenhouse gas emissions from global agricultural production: 1970–2050. *Global Change Biology, 22*(2), 763–781. https://doi.org/10.1111/gcb.13120

Climate Watch (2023). World Resource Institute Data, FAO FAOSTAT Agricultural (2020). Retrieved February 19, 2024, from https://www.fao.org/faostat/en/#data/QCL.

Dogan, N. (2016). Agriculture and Environmental Kuznets Curves in the case of Turkey: Evidence from the ARDL and Bounds test.

Dufour, J., Serrano, D. P., Gálvez, J. L., Moreno, J., & Garcia, C. (2009). Life cycle assessment of processes for hydrogen production, environmental feasibility and reduction of greenhouse gases emissions. *International Journal of Hydrogen Energy*, *34*(3), 1370–1376.

Erbas, B., & Solakoglu, E. (2017). In the presence of climate change, the use of fertilizers and the effect of income on agricultural emissions. *Sustainability*, *9*(11), 1989.

Fan, L., Niu, H., Yang, X., Qin, W., Bento, C. P., Ritsema, C. J., & Geissen, V. (2015). Factors affecting farmers' behaviour in pesticide use: Insights from a field study in northern China. *Science of the Total Environment*, *537*, 360–368.

FAO (2020). Emissions due to agriculture. Global, regional and country trends 2000–2018, FAOSTAT Analytical Brief Series No 18, Rome.

IPCC (2023). Summary for policymakers. In Climate Change 2023: Synthesis report. Contribution of Working Groups I, II and III to the sixth assessment report of the intergovernmental panel on climate change [Core Writing Team, H. Lee and J. Romero (Eds.)]. IPCC, Geneva, Switzerland, pp. 1–34. doi: 10.59327/IPCC/AR6-9789291691647.001

Liu, X., Zhang, S., & Bae, J. (2017). The nexus of renewable energy-agriculture-environment in BRICS. *Applied Energy*, *204*, 489–496.

Malhi, G. S., Kaur, M., & Kaushik, P. (2021). Impact of climate change on agriculture and its mitigation strategies: A review. *Sustainability*, *13*(3), 1318. https://doi.org/10.3390/su13031318

Mendelsohn, R. (2009). The impact of climate change on agriculture in developing countries. *Journal of Natural Resources Policy Research*, *1*(1), 5–19.

Moutinho, V., Madaleno, M., Inglesi-Lotz, R., & Dogan, E. (2018). Factors affecting CO_2 emissions in top countries on renewable energies: A LMDI decomposition application. *Renewable and Sustainable Energy Reviews*, *90*, 605–622.

Pata, U. K. (2021). Linking renewable energy, globalization, agriculture, CO_2 emissions and ecological footprint in BRIC countries: A sustainability perspective. *Renewable Energy*, *173*, 197–208. https://doi.org/10.1016/j.renene.2021.03.125

Pedroni, P. (1999). Critical values for cointegration tests in heterogeneous panels with multiple regressors. *Oxford Bulletin of Economics and Statistics, 61*(S1), 653–670.

Rehman, A., Ozturk, I., & Zhang, D. (2019). The causal connection between CO_2 emissions and agricultural productivity in Pakistan: Empirical evidence from an autoregressive distributed lag bounds testing approach. *Applied Sciences*, *9*(8), 1692.

Rojas-Downing, M. M., Nejadhashemi, A. P., Harrigan, T., & Woznicki, S. A. (2017). Climate change and livestock: Impacts, adaptation, and mitigation. *Climate Risk Management*, *16*, 145–163. https://doi.org/10.1016/j.crm.2017.02.001

Sarkodie, S., & Owusu, P. A. (2016). The relationship between carbon dioxide and agriculture in Ghana: A comparison of VECM and ARDL model. *Environmental Science and Pollution Research*, *23*, 10968–10982.

Schneider, U. A., & Smith, P. (2009). Energy intensities and greenhouse gas emission mitigation in global agriculture. *Energy Efficiency*, *2*, 195–206.

Škare, M., Tomić, D., & Stjepanović, S. (2020). Energy consumption and green GDP in Europe: A panel cointegration analysis 2008-2016. *Acta Montanistica Slovaca*, *25*, 1.

Ullah, S., Ahmad, W., Majeed, M. T., & Sohail, S. (2021). Asymmetric effects of premature deagriculturalization on economic growth and CO_2 emissions: Fresh evidence from Pakistan. *Environmental Science and Pollution Research*, *28*, 66772–66786.

Ullah, A., Khan, D., Khan, I., & Zheng, S. (2018). Does agricultural ecosystem cause environmental pollution in Pakistan? Promise and menace. *Environmental Science and Pollution Research*, *25*(14), 13938–13955. https://doi.org/10.1007/s11356-018-1530-4

Usman, A., Ozturk, I., Ullah, S., & Hassan, A. (2021). Does ICT have symmetric or asymmetric effects on CO_2 emissions? Evidence from selected Asian economies. *Technology in Society*, *67*, 101692.

Waheed, R., Chang, D., Sarwar, S., & Chen, W. (2018). Forest, agriculture, renewable energy, and CO_2 emission. *Journal of Cleaner Production, 172*, 4231–4238.

World Bank (2013). *World development indicators 2013*. World Bank Publications.

Xiao, L., Liu, J., & Ge, J. (2021). Dynamic game in agriculture and industry cross-sectoral water pollution governance in developing countries. *Agricultural Water Management*, *243*, 106417.

Zhangwei, L., & Xungangb, Z. (2011). Study on relationship between Sichuan agricultural carbon dioxide emissions and agricultural economic growth. *Energy Procedia, 5*, 1073–1077.

Chapter 13

Agricultural Production by Sector, Convergence and Environmental Sustainability

Ebikabowei Biedomo Aduku

Department of Economics, University of Africa, Toru-Orua, Bayelsa State, Nigeria

Abstract

This chapter specifically examines the convergence (divergence) effect of agricultural production by sector and the effect of agricultural production on environmental sustainability using West African countries as a case study. Annual time series variables covering the periods from 2000 to 2022 were empirically analysed using system dynamic panel-data estimation. Results showed significant beta-divergence in crop production, livestock production and total agricultural production, with a half-life of divergence rates of 10.71%, 13.15% and 16.91%, respectively, for crops, livestock and aggregate agricultural production. It was also found that the consumer prices index had a positive and significant effect on crop production, livestock production and total agricultural production. The findings also showed that crops and livestock production had a negative and significant effect on arable land and a positive and insignificant effect on CO_2 emissions. Also, total agricultural production had a negative and insignificant effect on arable land and a positive and insignificant effect on CO_2 emissions. It was also found that gross fixed capital formation had a negative and significant effect on arable land and a negative and insignificant effect on CO_2 emissions. The government and pertinent parties ought to promote sustainable farming practices such as conservation agriculture, rotational grazing, agroecology and reduced dependency on chemical inputs. Policies that support legislation that provide financial incentives

Sustainable Agricultural Practices: Economic and Environmental Implications, 191–207

Published under exclusive licence by Emerald Publishing Limited
doi:10.1108/978-1-83608-336-820241013

for environmentally sustainable agricultural activities and encourage of preservation of biodiversity are suggested.

Keywords: Agricultural production; convergence; dynamic panel; arable land; CO_2 emission; environmental sustainability

Introduction

There are four primary subsectors within the agricultural sector. These are the crops subsector, livestock (both production and animal health) subsector, fisheries and aquaculture (captures fisheries) subsector and the forestry subsector. Crops like wheat, cotton, rice, sugarcane and oilseed crops (sunflower, rapeseed, mustard, cottonseed, onion, chillies and potato) are included in the crops' subsector. The textile industry gets its raw materials from crops like cotton crops. Sugarcane crop is a cash crop and it is vital for sugar and sugar-related production, while wheat and rice are staple food crops; rice crop could serve as a source of foreign exchange earnings for any country. The livestock sub-sector plays a crucial part in economic development and it is also a source of foreign revenues. The domestic need for meat, milk and eggs is met by the livestock sub-sector. The livestock sub-sector plays a crucial role in reducing poverty and can uplift the socio-economic conditions of micro, small and medium-scale farmers and the landless rural poor. Enhancement of food security and the reduction of poverty are also among the major roles played by the fishery sub-sector. It is a vital source of export revenue, which is good for the economy. Forests are identified as essential components of the ecosystem, and their destruction could pose serious socioeconomic problems for future generations (Chandio et al., 2016).

No economy may function properly and sustainably without the growth that agriculture brings, as it is essential to both the process of economic growth and the fight against poverty. The majority of the poor, particularly those in emerging nations, rely on agriculture as their primary source of income and livelihood, making the sector one of the most dominant economic sectors. In nations like Nigeria, the agricultural sector accounts for over 70 per cent of the sectors that generate employment – with more than 36% of the labour force being employed (Oyaniran, 2020). Compared to other sectors, the agricultural sector growth, according to the World Bank (2023), has the potential to increase the earnings of the poorest people by two to four times higher. The agricultural sector is among the sectors that expedite economic growth and development (Oyaniran, 2020). This makes it possible to meet the needs of the growing population. It also releases excess labour to the industrial sector, improves domestic savings and consequently promotes domestic investment. Also, the agriculture sector either generates foreign exchange through exports or facilitates the saving of foreign exchange through import substitution. These, therefore, generate extra capital reserves that can be used to import technology for manufacturing production. In addition, by producing cheaper non-tradable agricultural items, the agriculture sector is known for contributing to the lowering of the wage-goods price, raising real wages and stimulating demand for domestic goods.

Agricultural growth could be linked to a reduction in the income gaps of countries through income convergence. Income convergence following the neoclassical theory means the tendency for the economy of less developed countries or poorer countries to grow faster than richer countries and achieve parity in per capita income with richer countries. Therefore, the income level of countries converges. Income in the current era has become one of the most common gauges for the standard of living. On this basis, it is proper to say that if income is converging, so is the standard of living of individuals in different countries, and the reverse if income is divergence. It, therefore, follows that countries that are experiencing income divergence are failing. Considering this, efforts are been made by countries and regions to be on the right side of income convergence. One deciding factor for income convergence of countries could lie in the level of agricultural growth by sector. An increase in agricultural growth by sector could offer enormous benefits in terms of value added across several production processes (Sertoğlu et al., 2017). This brings about dynamism and consolidation in sectors including synthetic fibres, plastics, as well as semiconductors among others. All of these participated in global markets. Thus, industrialization could be promoted including the redirection towards exports, capable of propelling demand for domestic activities connected to production overseas, creating more formal jobs and enabling savings to be reinvested in infrastructure and other resource-hungry areas of the economy. This demands the formation of a national technology system that can enable innovation growth in the production process among others. This could instigate higher productivity and value-added and will bring about higher income. However, an associated environmental issue could be environmental unsustainability.

Agriculture growth could result in deforestation as land could be expanded, implying the clearing of forests that could cause loss of habitat, decline in biodiversity and rise in greenhouse gas emission as carbon stored in trees and soil are released. Also, agricultural growth could come with intensive agricultural practices such as excessive tillage and heavy fertilizer chemical applications capable of causing soil degradation and could contribute to climate change among others (Food and Agriculture Organization of the United States, 2023). However, by applying more practices that are environmentally friendly, the impact on the environment could be reduced.

In recent decades, most emerging countries have grown faster than developed nations. Although this is not an unconditional convergence in which poorer countries grow faster than rich countries, poorer countries' growth rates have been much higher than those of developed countries. This may be likely, but its continuity is not guaranteed. Such astonishing evolution has important understudied implications for economic managers. There are compelling grounds to believe that countries that are poorer than the top richest countries of the day will expand faster than the leading ones. They may expand not only through the use of new technology but also by moving from the existing frontier to more contemporary technology. Incomes per capita increased in both developed and developing countries, but the growth rate in developed countries was typically higher. Developing countries continue to lag behind developed countries, and accompanying problems such as absolute poverty remains widespread (Ma et al., 2022; Martin, 2019).

Developing countries' growth accelerated in the 1990s, outpacing that of developed countries. This historic transformation attracts substantial interest from managers of the economy who are concerned with economic development; however, despite increased support for agricultural transformation, which might spark rapid growth in the agricultural sector, particularly in emerging nations, very little has been done. This support is reflected in national and regional development strategies and activities. It is safe to say that the recent agricultural transformation has made a vast part of the world, food secure, particularly in the African zone. Furthermore, with increasing agricultural output, African countries appear to be closing income gaps through income convergence (Martin, 2019).

However, at the same time, agricultural transformation may emit a large number of pollutants into the atmosphere, such as carbon dioxide, and methane among others, rendering it environmentally unsustainable. The empirical assessment in terms of its growth by sectors, convergence and environmental sustainability of the African countries is not widely reported in the literature. The specific objective of this chapter, therefore, is to analyse the convergence (divergence) effect of agricultural production by sector and to examine the impact of agricultural production on environmental sustainability. The chapter will be of great importance to governments of countries mostly in Africa, stakeholders and academics. The chapter will also serve as a good area of discovery for further research.

Extant Literature

Agriculture is the raising of plants and animals for human use. According to the National Geographic Society (2024), agriculture is the art and science of nourishing the earth, raising crops and producing animals for food and other purposes including economic gain. Harris and Fuller (2014) provided a similar definition, defining agriculture as the art and science of tilling the land, cultivating crops and rearing livestock. Rubenstein (2003) viewed the concept as a deliberate attempt to restructure a section of the earth's surface by growing crops and rearing animals for human use or commercial benefit. Agriculture can also be viewed as the deliberate human-managed cultivation of livestock and beneficial plants. The agricultural subsectors considered, for this chapter, are the crops sector and the livestock sector.

The area of agriculture that deals with growing crops for food and fibre is called crop production. It comprises all of the feed sources needed to sustain and grow crops. Producing and managing row crops for commercial purposes, such as soybean, maize, cotton, wheat and tobacco, is referred to as crop production. It can also be described as the crop agricultural subsector output. On the other hand, farming raising terrestrial animals for food, fibre and labour is known as the livestock sector. The sector includes meat and milk from all sources, dairy products such as cheese, and eggs, honey, raw silk, wool, and hides and skins.

The domestic need for meat, milk and eggs is satisfied by this industry. It also contributes to the generation of jobs and foreign exchange.

Convergence is among the stylized facts of economic growth that the neoclassical growth models attempted to predict. It suggests that there is a negative relationship between initial levels of income and the rates of income growth at a given time, contingent on factors important for economic growth. This implies that provided certain criteria are met, rich countries typically expand slower than impoverished countries. The neoclassical growth models described this type of convergence as conditional convergence and accurately predict this type of convergence where economies share comparable economic characteristics or technology and preferences. Neoclassical growth models are predicated on an economy having a utility function that expresses its preferences and a particular production function. The economy will eventually reach what is known as the steady-state equilibrium, where it will no longer be able to grow, based on certain assumptions. There is a divergence when the economy is moving away from its steady state and convergence when it is getting closer to it. The utilization of neoclassical growth models allows for the computation of the rate of convergence at which the economy approaches its steady state (Quiroga, 2007). Convergence implies that the economic measures or variables of countries gradually approach and catch up with themselves and move towards a stable state. This chapter, therefore, aims to further investigate if agricultural production by sector shows this trend.

One of the issues closely linked with growth in agricultural production is environmental sustainability. The appropriate use and management of natural resources to guarantee their availability for present and future generations while minimizing adverse effects on the environment is referred to as environmental sustainability. It entails providing for current needs without sacrificing the capacity of future generations to provide for themselves (Goodland, 1995; World Economic and Social Survey, 2015). Among the common indicators of environmental sustainability are arable land (% of land area) and carbon dioxide emissions (CO_2 emissions). Arable land includes land defined by the Food and Agriculture Organization (FAO) as land under temporary crops (double-cropped areas are counted once), temporary meadows for mowing or pasture, land under market or kitchen gardens, and land temporarily fallow. Land abandoned as a result of shifting cultivation is excluded (World Bank, 2024). CO_2 emissions, on the other hand, are those stemming from the burning of fossil fuels and manufacturing activities, among others. They include CO_2 produced during consumption of solid, liquid and gas fuels and gas flaring, among others (World Bank, 2024).

In the empirical font, a cross-sectoral approach was used by Moreno-Miranda and Dries (2022) to evaluate and contrast the sustainability performance of Ecuadorian fruit and vegetable farms. The study also examines the relationship between farm resources, farmer demographics and farm size on the one hand, and sustainability outcomes on the other. The data was analysed using general linear regression and principal component analysis. It was found that farms with superior social and economic sustainability performance were better at using traditional resources (labour and land). Also, farms with a higher environmental-oriented performance were found to be more environmentally sustainable. With an application to Bangladesh, Li et al. (2021) investigated the possibility of enhancing the economic-environmental performance of crop production using spatially integrated modelling and optimization. The findings indicated that accounting for crop-specific, seasonal and geographical differences in crop nitrogen-use efficiency and nitrogen transport would result in an 83% reduction in the per-billion-taka nitrogen loss from the soil compared to the baseline level. Lankoski and Lankoski (2023) use necessary condition analysis to identify probable necessary conditions (bottlenecks) and determine the necessary levels of those circumstances for both strong and weak environmental sustainability outcomes in agriculture. The findings demonstrated that low levels of cropland share, livestock density and production intensity were prerequisites for high levels of environmental sustainability. These in turn are impacted by agricultural support policies, both historical and current, that are production-coupled and limit the attainment of high environmental sustainability. These policies include market price support and output payments. Also, in North-Central Nigeria, Tsue et al. (2014) investigated the impact of arable land ownership and use on environmental sustainability. A sample of 356 respondents was used for the study. The data was analysed using descriptive and inferential statistics. It was found that the improvement in environmental sustainability was significantly impacted by land tenure security.

The literature on environmental sustainability is enormous, but fewer studies have empirically examined the relationship between agricultural production and environmental sustainability. In the West African subregion, empirical evidence on agricultural production and convergence in particular, and agricultural production and environmental sustainability especially at the panel level is missing. This chapter, which examines the convergence (divergence) effect of agricultural production by sector, the divergence rate, the half-life of divergence and the effect of agricultural production on environmental sustainability is, therefore, timely and appropriate. The contributions of this chapter to the literature will be useful for agricultural and environmental sustainability planning and thus will be of great importance to governments of countries mostly in Africa, stakeholders and academics.

Methodology

Variables and Data Source

The variables are agricultural growth by sector – measured by the crop production index, livestock production index and total agricultural production index per capita; consumer price index; and total government expenditure (% of GDP), which measures the extent of government fiscal policy. Other variables include environmental sustainability – measured by arable land (% of land area) and CO_2 emissions (metric tons per capita); GDP per capita; gross fixed capital formation (% of GDP) – measure for domestic investment and population growth (annual %) – proxy for labour force growth. The data is a time series data that covers the periods from 2000 to 2022. The data is sourced from the World Development

Indicator – WDI open data portal and the African Development Bank Socio-Economic Database. Consumer price index, arable land (% of land area), GDP per capita, gross fixed capital formation (% of GDP) and population growth (annual %) are sourced from the WDI open data portal, while the data for the crop production index, livestock production index, total agricultural production index per capita, CO2 emissions (metric tons per capita), and total government expenditure (% of GDP) are sourced from the African Development Bank Socio-Economic Database.

Empirical Model

To analyse the convergence or divergence effect of agricultural growth by sector, the following functional specification is used:

$$\mathrm{AGS}_{it} = f\left(\mathrm{CPI}_{it}\ \mathrm{GXA}_{it}, \mathrm{LAB}\right) \tag{1}$$

where AGS is agricultural production by sector – measured by the crop production index, livestock production index and total agricultural production index per capita. CPI is the consumer price index, GXA is total government expenditure (% of GDP), which measures the extent of government fiscal policy on agriculture, and LAB is population growth (annual %) – a proxy for labour force growth. CPI, GXA and LAB in equation (1) are conditioning variables of country i in t years augmented into the convergence model. The conditional beta-convergence regression model is

$$\mathrm{AGS}_{it} = a + \beta \mathrm{AGS}_{it-1} + \varphi_1 \mathrm{CPI}_{it} + \varphi_2 \mathrm{GXA}_{it} + \varphi_3 \mathrm{LAB}_{it} + \gamma_{it} + \varepsilon_{it} \tag{2}$$

where γ_{it} measures a country-specific effect and ε_{it} is an error term. The beta-convergence is represented by the coefficient β. A negative value of β means that there is beta-convergence, while a positive value implies that the agricultural growth levels of economies are diverging (beta-divergence). Convergence means that the agricultural growth levels of countries within the West African region become more similar over time, indicating that countries in the region with low-sectoral agricultural growth are growing faster in agricultural sector output compared to economies with high-sectoral agricultural growth. While divergence means that the agricultural production levels of countries within the West African region become more dissimilar over time, indicating that economies in the region with high-sectoral agricultural growth are growing faster in terms of agricultural sector output than those with low-sectoral agricultural growth. The convergence rate and the half-life cycle of convergence for the inquiry time, following Li et al. (2023) can be computed using the following formulas:

$$\mathrm{CR} = -\ln\left(1 - |\beta|\right) / t$$
$$\mathrm{HL} = (\ln 2) / CR$$

where CR indicates the convergence rate within the period t of the research and HL is the half-life of convergence. In the West African region, most of the economies are at similar economic and institutional levels of development. Therefore, it is expected that there will be a convergence effect. If that is the case, then we say that countries in the region with low-sectoral agricultural growth are 'catching up' with countries in the region with high sectoral agricultural growth.

On the other hand, to examine the impact of agricultural production on environmental sustainability, the following functional specification is used:

$$\text{NSUB} = f(\text{CRP}, \text{LSP}, \text{TAP}, \text{PGDP}, \text{GFCF}, \text{LABs}) \tag{3}$$

where ENSUB is environmental sustainability – measured by arable land (% of land area), and CO_2 emissions (metric tons per capita). CRP is crop production – measured by the crop production index, LSP is livestock production – measured by the livestock production index, and TAP is total agricultural production – measured by the total agricultural production index per capita. PGDP represents GDP per capita, while GFCF is gross fixed capital formation (% of GDP), a measure for domestic investment, and LAB is population growth (annual %), a proxy for labour force growth. The panel model specification of equation (3) is

$$\begin{aligned}\text{ENSUB}_{it} = {} & \vartheta_0 + \vartheta_1\text{CRP}_{it} + \vartheta_2\text{LSP}_{it} + \vartheta_3\text{TAP}_{it} + \vartheta_4\text{PGDP}_{it} + \vartheta_5\text{GFCF}_{it} \\ & + \vartheta_6\text{LAB}_{it} + \gamma_{it} + \varepsilon_{it}\end{aligned} \tag{4}$$

where γ_{it} is a between-country error term, ε_{it} is a within-country error term, i is observational units, while t is period. To specify a linear model with a dynamic variable ($\text{ENSUB}_{i,t-1}$), we include the dynamic variable as

$$\begin{aligned}\text{ENSUB}_{it} = {} & \vartheta_0 + \vartheta_1\text{ENSUB}_{i,t-1} + \vartheta_2\text{CRP}_{it} + \vartheta_3\text{LSP}_{it} + \vartheta_4\text{TAP}_{it} + \vartheta_5\text{PGDP}_{it} \\ & + \vartheta_6\text{GFCF}_{it} + \vartheta_7\text{LAB}_{it} + \gamma_{it} + \varepsilon_{it}\end{aligned} \tag{5}$$

Where all variables remained as defined earlier. It is well documented in the literature that initial values of the dependent variable are often correlated with the lagged dependent variable appearing as an explanatory variable, therefore, leading to an endogeneity problem. Equation (2) – the convergence equation as well as equation (5) – the environmental sustainability equation is not an exception. The inclusion of AGS_{it-1} and $\text{ENSUB}_{i,t-1}$ could cause endogeneity. Fixed effects, such as geography and demography, can be associated with the explanatory variables. The error term captures fixed effects, both unobserved country-specific effects (γ_{it}) and observation-specific errors (ε_{it}).

The equations could be estimated using two-stage least squares (2SLS). However, at the first stage of the two-stage least square (2SLS) regression, the instruments may be weeks, which could make the technique biased, similar to the ordinary least square technique. Also, the fixed effect and random effect estimators may be inconsistent because of the presence of the unobserved country-specific effects (γ_{it}) in the equation for estimation. As a result, the generalised method of moments (GMM) estimator will be used as the estimation technique. The

GMM estimator is widely acknowledged for its ability to account for both endogeneity difficulties and unobserved country-specific effects. As a result, GMM approaches have become frequently used. In the difference equation, the system GMM approach employs the lagged values of endogenous variables, while in the level equation, the initial difference is used (Ali, 2017; Ugbor et al., 2018). When using instrument-based estimate techniques, it is common to assess their validity (exogeneity). The Hansen test is employed to test for the validity of instruments.

Results

Unit Root Test

The time series variables are tested for unit roots by conventional practice. The Fisher-type unit-root test and Im, Pesaran and Shin (IPS) unit root tests are used. Also, the cointegration test is conducted using the Pedroni test to determine the presence of a long-run relationship among the variables in the models for objectives one and two. The result is reported in Table 13.1.

The insignificant test statistics at the level in both the Fisher-type unit-root test and the IPS test suggest the acceptance of the null hypothesis that all panels contain unit roots. Therefore, the variables were differenced once and the test was repeated with the differenced series using both the Fisher-type unit-root test and the IPS test. At 1st difference, the test statistics in both tests are significant. Therefore, this suggests the rejection of the null hypothesis at 1st difference. This means that at least one panel is stationary at 1st difference. To ascertain the long-term relationship between the variables in the model, a cointegration test is carried out. The Pedroni test for cointegration is employed, and the test result is also reported in Table 13.1.

The Phillips–Perron *t* and Augmented Dickey−Fuller *t* statistics are significant at the 5% level. Therefore, rejecting the null hypothesis of no cointegration. This tells us that the variables in the respective equations for objectives one and two are cointegrated in all panels, with a panel-specific cointegrating vector. This also means that the variables have a significant relationship in the long run. In the long run, the variables return to equilibrium in the event of a shock or disparity.

Convergence (Divergence) Effect of Agricultural Production by Sector and the Effect of Agricultural Production on Environmental Sustainability

Convergence (Divergence) Effect of Agricultural Production by Sector

The first objective of the chapter is to determine the convergence (divergence) effect of agricultural production by sector. Equation (2) was, therefore, estimated using the system dynamic panel-data GMM estimation, and the convergence rate and the half-life of convergence were computed within the period of the study. The result is presented in panel A in Table 13.2. Panel A is divided into three columns. Column (1) reports the results for the crops agricultural subsector; in column (2), the results for livestock agricultural subsector are presented, while the results for total agricultural production are reported in column (3).

Table 13.1. Unit Root and Cointegration Test Results.

Variable	**Fisher-Type Unit-Root Test Test Result**		**Im, Persaran and Shin Unit-root Test Result**		**~*I*(*d*)**
	Level	**1st Difference**	**Level**	**1st Difference**	
TAP	−2.1223 (0.0169)	−18.0755* (0.0000)	−1.8521 (0.0320)	−10.2592* (0.0000)	*I*(1)
CPR	−0.5488 (0.2916)	−20.2320* (0.0000)	−0.2775 (0.3907)	−11.0535* (0.0000)	*I*(1)
LSP	−3.6695 (0.9999)	−13.2879* (0.0000)	−1.7824 (0.9999)	−8.6563* (0.0000)	*I*(1)
CPI	−3.4163 (0.9997)	−12.8271* (0.0000)	−0.7049 (0.9998)	−8.4561* (0.0000)	*I*(1)
GXA	−3.0610 (0.7011)	−18.4240* (0.0000)	−0.7467 (0.1030)	−10.4394* (0.0000)	*I*(1)
PGDP	−2.1496 (0.9842)	−11.3212* (0.0000)	−2.0240 (−0.9785)	−7.6409* (0.0000)	*I*(1)
GFCF	−1.0966 (0.8636)	−13.6358* (0.0000)	0.9895 (0.8388)	−8.6487* (0.0000)	*I*(1)
LAB	−2.5371 (0.0756)	−12.3214* (0.0000)	−0.0293 (0.1212)	−8.0988* (0.0000)	*I*(1)
ALAND	−1.4162 (0.0756)	−9.9418* (0.0000)	−1.3866 (0.0828)	−6.8090* (0.0000)	*I*(1)
CO_2	−0.3239 (0.3730)	−13.5623* (0.0000)	−0.6030 (0.2733)	−8.9186* (0.0000)	*I*(1)

The Fisher-type unit-root test (based on augmented Dickey−Fuller tests), and Im, Pesaran and Shin (IPS) ***P***-values are presented in parentheses.

Results of Pedroni Test for Cointegration

For the Convergence Equation			For the Environmental Sustainability Equation		
	Statistic	*P*-value		Statistic	*P*-value
Modified Phillips–Perron ***t***	0.4283	0.3342	Modified Phillips–Perron t	4.2127	0.0000
Phillips–Perron ***t***	−1.9822	0.0237	Phillips–Perron t	−1.6208	0.0525
Augmented Dickey−Fuller ***t***	−2.2006	0.0139	Augmented Dickey−Fuller *t*	−2.6025	0.0046

Source: Computed by the author.

Table 13.2. Estimates of the Convergence (Divergence) Effect of Agricultural Production by Sector, and the Impact of Agricultural Production on Environmental Sustainability.

Panel A: Results for objective one – the convergence (divergence) effect of agricultural production by sector			
	(1) Crops production	**(2) Livestock production**	**(3) Total agricultural production**
CRP_{t-1}; LSP_{t-1}; TAP_{t-1}	0.7745	0.7037	0.6106
	(z = 21.90; P = 0.000)	(z = 22.98; P = 0.000)	(z = 15.61; P = 0.000)
CPI	0.4667	0.9144	0.0429
	(z = 4.15; P = 0.000)	(z = 40.73; P = 0.000)	(z = 2.43; P = 0.008)
GXA	−0.0330	−0.0995	0.0923
	(z = −0.19; P = 0.848)	(z = −2.74; P = 0.006)	(z = 0.59; P = 0.558)
LAB	1.0613	0.0008	1.8830
	(z = 0.77; P = 0.442)	(z = 0.34; P = 0.736)	(z = 1.46; P = 0.146)
Constant	−12.4326	1.1789	26.6962
	(z = −2.09; P = 0.037)	(z = 0.99; P = 0.321)	(z = 4.57; P = 0.000)
Wald χ^2 (9)	1871.11 (P = 0.0000)	22144.64 (P = 0.0000)	278.02 (P = 0.0000)
Hansen test	355.0023 (P = 0.0463)	343.3815 (P = 0.0169)	351.6297 (P = 0.0069)
Divergence rate	0.0281	0.0229	0.0178
Half-life of divergence	10.7128	13.1454	16.9118

(Continued)

Table 13.2. (*Continued*)

Panel B: Results for objective two – effect of agricultural production on environmental sustainability		
	(1) **Arable land (% of land area)**	**(2)** **CO_2 emissions (metric tons per capita)**
LAB	−0.0820 (z = −0.50; P = 0.616)	0.0081 (z = 0.54; P = 0.586)
CRP	−0.0152 (z = −6.34; P = 0.000)	0.0001 (z = 0.29; P = 0.774)
LSP	−0.0222 (z = −5.25; P = 0.000)	0.0005 (z = 1.31; P = 0.189)
TAP	−0.0003 (z = −0.11; P = 0.912)	0.0008 (z = 3.05; P = 0.002)
PGDP	0.0001 (z = 0.48; P = 0.629)	0.0001 (z = 1.22; P = 0.223)
GFCF	−0.0153 (z = −2.03; P = 0.042)	−0.0003 (z = −0.52; P = 0.603)
Constant	−0.0133 (z = −0.02; P = 0.981)	0.0151 (z = 0.34; P = 0.735)
Wald χ^2 (9)	11,923.50 (P = 0.0000)	1,250.58 (P = 0.0000)
Sargan test	429.1827 (P = 0.0000)	330.092 (P = 0.0081)

Source: Computed by the author.

The beta coefficients (βs) for crop production, livestock production and total agricultural production are 0.7745, 0.7037 and 0.6106 with significant z-values of 21.90, 22.98 and 15.61, respectively. The positive and significant beta-coefficients mean significant beta-divergence in crop production, livestock production and total agricultural production. The divergence rates and the half-life of divergence rates for crop production are 0.0281 and 10.7128, livestock production is 0.0229 and 13.1454 and total agricultural production is 0.0178 and 16.9118, respectively.

As regards the conditioning variables, the consumer price index coefficient is positive and significant in columns (1), (2) and (3). This means that the consumer prices index has a positive and significant effect on crop production, livestock production and total agricultural production. Specifically, a percentage increase in the consumer price index results in a significant increase in crop production, livestock production and total agricultural production by 0.47%, 0.91% and 0.04%, respectively.

Total government expenditure (% of GDP) has a negative and insignificant effect on crop production in column (1), a negative and significant effect on livestock production in column (2), and a positive and insignificant effect on total agricultural production in column (3).

The coefficient for population growth (annual %) is positive and insignificant in columns (1), (2) and (3). This means that population growth (annual %) has a positive and insignificant effect on crop production, livestock production and total agricultural production.

The Wald χ^2 (9) statistics are 1,871.11 (P = 0.0000), 22,144.64 (P = 0.0000) and 278.02 (P = 0.0000), respectively, for columns (1), (2) and (3). The significant P-values suggest rejecting the null hypothesis that the variables jointly have no significant effect on crop production, livestock production and total agricultural production. Therefore, the variables have significant joint effects on crop production, livestock production and total agricultural production. Also, the significant results of the Sargan test of overidentifying restrictions, which was used to assess the validity (exogeneity) of the instruments, were obtained, consequently, verifying the instruments'' validity (exogeneity).

Effect of Agricultural Production on Environmental Sustainability

In line with the second objective, the effect of agricultural production on environmental sustainability was estimated using the system dynamic panel-data GMM estimation, and the result is presented in panel B in Table 13.2. Panel B is divided into two columns. Column (1) reports the results for arable land (% of land area), while the results for CO_2 emissions (metric tons per capita) are reported in column (3).

The coefficients for population growth (annual %) are −0.0820 and 0.0081 in columns (1) and (2) with a z-value of −0.50 and 0.54. This means that the population growth (annual %) has a negative and insignificant effect on the arable land (% of land area) and a positive and significant effect on CO_2 emissions (metric tons per capita).

Crop production showed a negative and significant coefficient of −0.0152 with a z-value of −6.34 in column (1) and a positive coefficient of 0.0001 with an insignificant *z*-value of 0.29 in column (2). This means that crop production has a negative and significant effect on arable land (% of land area) and a positive and insignificant effect on CO_2 emissions (metric tons per capita).

A similar result showed up for livestock production. Livestock production has a negative and significant effect on arable land (% of land area) in column (1) and a positive and insignificant effect on CO_2 emissions (metric tons per capita) in column (2).

As regards total agricultural production, the result showed a negative and insignificant coefficient in column (1) and a positive and significant coefficient in column (2). This means that total agricultural production has a negative and insignificant effect on arable land (% of land area) and a positive and insignificant effect on CO_2 emissions (metric tons per capita).

Per capita GDP is positive and insignificant in both columns (1) and (2). This means that the effect of per capita GDP on arable land (% of land area) is positive and insignificant. Similarly, the effect of per capita GDP on CO_2 emissions (metric tons per capita) is positive and insignificant.

The gross fixed capital formation showed a coefficient of −0.0153 and a *z*-value of −2.03 in column (1). In column(s), the coefficient is −0.0003 with a *z*-value of −0.52. This means that gross fixed capital formation has a negative and significant effect on arable land (% of land area) and a negative and insignificant effect on CO_2 emissions (metric tons per capita).

The Wald χ^2 (9) statistics are 11,923.50 ($P = 0.0000$) and 1,250.58 ($P = 0.0000$) for columns (1) and (2). The significant *P*-values suggest rejecting the null hypothesis that the variables jointly have no significant effect on arable land (% of land area) and CO_2 emissions (metric tons per capita). Therefore, the variables have a significant joint effect on arable land (% of land area) and CO2 emissions (metric tons per capita). Also, the significant results of the Sargan test of overidentifying restrictions, which was used to assess the validity (exogeneity) of the instruments, were obtained. Therefore, confirming the instruments' validity (exogeneity).

Discussion

The positive and significant beta-coefficients mean significant beta-divergence in crop production, livestock production and total agricultural production. This tells us that the agricultural growth levels of economies are diverging. That is the agricultural production levels of countries within the West African region are more dissimilar within the study period indicating that economies in the region with high sectoral agricultural growth are growing faster in terms of agricultural sectoral production than those with low-sectoral agricultural growth. Specifically, for the crops subsector, the rate at which economies in the region diversify in crops subsector agricultural production is 0.03% per annum, while the divergence rate for livestock production is 0.02% per annum. The rate at which total agricultural production diverse among countries of high and low agricultural production is about 0.02%. The half-life of divergence estimates revealed that it would take about 10 years and 7 months for economies in the region with high crop production to reduce to half of its current level of crop production and reduce to half of the current rate of divergence. For the livestock subsector, it will take about 13 years and 1 month for economies in the region with high livestock production to reduce to half of its current level of production and reduce to half of the current rate of divergence of the sector's production. As regards total agricultural production, it will take about 16 years and 9 months for economies in the region with high total agricultural production to reduce to half of its current level of production and reduce to half of the current rate of divergence.

An increase in agricultural production has a detrimental effect on the environment; it reduces environmental sustainability. For example, an increase in both crop production and livestock production contributes significantly to a reduction in arable land, while total agricultural production contributes significantly to an increase in CO_2 emissions. This could be attributed to a high level of unsustainable agricultural practices including monoculture, excessive application of chemical inputs, practice of conventional tillage, overgrazing, etc. Sustainable farming methods including agroecology, conservation agriculture, rotational grazing and less reliance on chemical inputs are needed to reduce the effect on the environment. Policy focusing on promoting the preservation of biodiversity and endorsing laws that provide financial incentives for ecologically friendly agricultural operations could be appropriate.

Other findings informed us that the consumer price index significantly instigates crops, livestock and total agricultural production. Government expenditure, on the other hand, has a negative effect, which could be linked to the low and inadequate agricultural financing by economies in the region. Population growth does not contribute significantly to change in agricultural production. Economic development of the region also reduces environmental sustainability. This means that the economies in the region have not developed to a level of influencing environmental sustainability.

Conclusion and Recommendations

This chapter has examined agricultural production by sector, convergence and environmental sustainability using West African countries as a case study. Annual time series variables covering the periods from 2000 to 2022 were empirically analysed using system dynamic panel-data estimation. Based on the findings, it is concluded that there is a divergence effect of agricultural production by sector. The agricultural growth levels of economies are diverging, at a rate of 0.03%, 0.02% and 0.02%, respectively, for the crops subsector, livestock subsector and total agricultural production. Reducing crop production to half of its current level and halving the current pace of divergence will take roughly 10 years and 7 months. While it will take roughly 13 years and 1 month for the livestock subsector and about 16 years and 9 months for total agricultural production to reduce to half of its current level and halve the current pace of divergence. Also, there is a significant effect of agricultural production on environmental sustainability. While total agricultural production greatly increases CO_2 emissions, increases in both livestock and crop productivity significantly reduce the amount of arable land available for use.

The government and pertinent parties ought to promote sustainable farming practices such as conservation agriculture, rotational grazing, agroecology and reduced dependency on chemical inputs. Policies that support legislation that provide financial incentives for environmentally sustainable agricultural activities and encourage of preservation of biodiversity are suggested.

References

Ali, S. (2017). Exchange rate effects on agricultural exports: Firm-level evidence from Pakistan. In University of Nottingham Discussion Paper No. 09.

Chandio, A. A., Yuansheng, J., & Magsi, H. (2016). Agricultural sub-sectors performance: An analysis of sector-wise share in agriculture GDP of Pakistan. *International Journal of Economics and Finance*, *8*(2), 156–162.

Erenstein, O. (2003). Smallholder conservation farming in the tropics and sub-tropics: A guide to the development and dissemination of mulching with crop residues and cover crops. *Agriculture, Ecosystems and Environment*, *100*, 17–37.

Food and Agriculture Organization of the United State (2023). Environmental sustainability in agriculture. https://www.fao.org/3/cc9065en/cc9065en.pdf

Goodland, R. (1995). The concept of environmental sustainability. *Annual Review of Ecology and Systems*, *26*, 1–24.

Harris, D. R., & Fuller, D. Q. (2014). *Agriculture: Definition and overview*. *Encyclopedia of Global Archaeology* (pp. 104–113). Springer.

Lankoski, J., & Lankoski, L. (2023). Environmental sustainability in agriculture: Identification of bottlenecks. *Ecological Economies*, *204*(A), 106–122.

Li, M., Guo, Z., & Zhang, W. (2021). Balancing food security and environmental sustainability by optimizing seasonal-spatial crop production in Bangladesh. *Environmental Research Letters*, *16*, 074046. https://iopscience.iop.org/article/10.1088/1748-9326/ac0be4/pdf

Ma, L., Zhang, Y., Chen, S., Yu, L., & Zhu, Y. (2022). Environmental effects and their causes of agricultural production: Evidence from the farming regions of China. *Ecological Indicators*, *144*, 24–46.

Martin, W. (2019). Economic growth, convergence, and agricultural economics. *Agricultural Economics*, *50*, 7–27.

Moreno-Miranda, C., & Dries, L. (2022). Assessing the sustainability of agricultural production – A cross-sectoral comparison of the blackberry, tomato and tree tomato sectors in Ecuador. *International Journal of Agricultural Sustainability*, *20*(7), 1373–1396.

National Geographic Society (2024). The art and science of agriculture. https://education.nationalgeographic.org/resource/the-art-and-science-of-agriculture/

Oyaniran, T. (2020). Current state of Nigeria agriculture and agribusiness sector. A paper presented at the African Continental Free Trade Agreement (AfCFTA) Workshop, September. https://www.pwc.com/ng/en/assets/pdf/afcfta-agribusiness-current-state-nigeria-agriculture-sector.pdf

Quiroga, P. A. B. (2007). History and evidence of economic convergence in Latin America. Instituto de Estudios Avanzados en Desarrollo (INESAD). https://www.jstor.org/stable/pdf/resrep00574.4.pdf

Rubenstein, J. M. (2003). *The cultural landscape: An introduction to human geography* (7th ed., p. 496). Pearson Education, Inc.

Sertoğlu, K., Ugural, S., & Bekun, F. V. (2017). The contribution of agricultural sector on economic growth of Nigeria. *International Journal of Economics and Financial Issues*, *7*(1), 547–552.

Tsue, P. T., Nweze, N. J., & Okoye, C. U. (2014). Effects of arable land tenure and use on environmental sustainability in north-central Nigeria. *Journal of Agriculture and Sustainability*, *6*(1), 14–38.

Ugbor, I. K., Ugbor, U. J., & Aduku, E. (2018). Governance structure and economic diversification. In *1st international conference of the Department of Economics*, University of Nigeria, Nsukka.

World Bank (2024). World Bank open data. https://data.worldbank.org/

World Bank (2023). Agriculture and food. https://www.worldbank.org/en/topic/agriculture/overview

World Economic and Social Survey (2015). Environmental sustainability. https://www.un.org/development/desa/dpad/wp-content/uploads/sites/45/publication/2015wess_ch5_en.pdf

Chapter 14

Ways for Enhancing Contribution of Architecture Engineering and Construction Industry and its Professionals to Food Security and Sustainable and Resilient Agricultural Practices

Begum Sertyesilisik

Istanbul University, Türkiye

Abstract

As technology aspect of agriculture becomes more and more important with the time to increase agricultural productivity in a sustainable and smart way, agriculture practices become more interdisciplinary. Furthermore, agricultural practices are affected by urban and rural planning enabling urban and rural farming. Architecture, engineering and construction (AEC) industry can support food security through the integration of agricultural practices and technologies into the built environment, its interior design, and greenhouses supporting urban and rural farming. Based on the literature review, this chapter aims to investigate ways for enhancing AEC industry's and its professionals' contribution to food security and sustainable agricultural practices. This chapter highlights roles of the AEC industry in enhancing food security and sustainable agricultural practices. This chapter emphasizes the importance of undergraduate and graduate curriculums of future AEC industry professionals (e.g., architects, interior architects, civil engineers) to equip them with the skills and knowledge of sustainable agricultural practices and technologies integrated greenhouses, built environment and indoor environment, and interior design. For this reason, agricultural policies need to cover food security-related interdisciplinary education and training (e.g., renewable energy-based agriculture integrated built environment) of AEC industry professionals. Agricultural

Sustainable Agricultural Practices: Economic and Environmental Implications, 209–222

Published under exclusive licence by Emerald Publishing Limited
doi:10.1108/978-1-83608-336-820241014

policies need to be designed with the contribution of and considering AEC industry professionals as they are among the main stakeholders of food security and renewable energy-based agriculture-integrated built environment. Furthermore, this chapter highlights how AEC industry, in compliance with United Nations Sustainable Development Goals and countries sustainable and resilient development plans, can contribute to food security and sustainability. This chapter can be beneficial to all stakeholders of the sustainable agricultural practices.

Keywords: Construction industry; food security; urban farming; urban planning; rural farming; rural planning; sustainable agricultural practices; sustainable agricultural technologies; AEC industry professionals; indoor environment design; sustainable built environment; interior design

Introduction

Healthy agricultural outputs and food security are at the centre of the sustainable and resilient development as they are key to the health and well-being of today's generation and next generation. Food security is the foundation of country's security, economic development through supporting different industries (e.g., agriculture, tourism) and economic growth, attracted investment and social stability (through reduced poverty, inequality and conflict) (Lee et al., 2024). There have been many attempts to solve the food insecurity problem. For example, global food security increased in the Green Revolution period (1960–2000) when global food production increased due to intensified rural agriculture (Armanda et al., 2019).

Food insecurity continues to be a worldwide threat (Meliko et al., 2023). World is not on track of achieving the SDG2 and the 2030 plan and humanity encounters challenges related with availability, affordability and accessibility of food in villages and cities (Dewan et al., 2024). Global Food Security Index's 11th edition indicated that reduced food security, vulnerability of the deteriorated food environment, reduced affordability of food and inequity in the global food system due to widened food security gap among the top performing and the worst performing countries in the world are observed (Economist website, n.d.).

Climate change has effects on food security and increased uncertainty risks (Lee et al., 2024), reduced yields (Challinor et al., 2014 as cited in Dewan et al., 2024), and changes in distribution and behaviour of species (Dewan et al., 2024). Lee et al.'s (2024) research on China indicated that climate change causes regional heterogeneity with respect to food security. Furthermore, air pollution and depleted ozone layer adversely affect the agricultural outputs (Dewan et al., 2024).

Many challenges can influence agriculture practices, food security and resilient and sustainable agriculture facilities in urban and rural areas. Integrated thinking of cities and villages (Sertyesilisik, 2023), and sustainable agriculture and environment policies for food security (Sertyesilisik, 2022) are important. Cities encounter challenges related with food security (Filippini et al., 2019).

Urbanization's inability to ensure food security is among its main problems (Abebe, 2024). Intensified innovative urban agriculture (UA) can have potential to support global food security (Armanda et al., 2019). As technology aspect of agriculture becomes increasingly important to increase agricultural productivity in a sustainable and smart way, agriculture practices become more interdisciplinary. Architecture-engineering and construction industry (AEC) and its expert professionals are among the main constructs and stakeholders of agricultural practices and facilities which are affected by urban and rural planning and built environment. Furthermore, agricultural facilities and UA rely on the outputs of the AEC industry and its experts. This industry can support and influence food security through the integration of agricultural practices and technologies into the built environment and greenhouses supporting urban and rural farming. Integrated thinking of cities and villages can create synergy (Sertyesilisik, 2023).

With the background that the AEC industry is vital for the construction of facilities/buildings (e.g., healthy and resilient food storage buildings) necessary to eliminate/minimize food waste, this chapter aims to investigate ways for enhancing AEC industry's and its professionals' contribution to food security and sustainable agricultural practices.

Food Security Challenges, Urbanization and Urban Agriculture

Urbanization and its challenges can have influence on food security. Abebe's (2024) research on Ethiopia revealed urbanization's effect on food security and emphasized that in case of rapid urbanization, production is reduced and food prices are increased. Furthermore, cropland and food security can be affected by the urbanization caused construction land expansion to cropland (Li et al., 2023). Additionally, migration from rural to urban areas and increased urban population concentration can contribute to cities' vulnerability challenging food security in future cities (de Amorim et al., 2019). Informal settlements can further challenge urbanization. For example, informal settlements in South Africa encounter challenges related with food insecurity (e.g., malnutrition) (Mkhize et al., 2023).

Many studies have emphasized UA's role in and potential for food security. For example, Abebe's (2024) research emphasized that food security can be enabled by land use planning and policies, widespread of UA and intensified farm. Pueyo-Ros et al.'s (2024) research supported UA's contribution to urban sustainability. Furthermore, Badami and Ramankutty's (2015) research revealed UA's feasibility in high-income countries. Similarly, Rezai et al. (2016) indicated that UA (in countries such as Malaysia) can contribute to food security through enabling citizens to access food. Additionally, Vieira and Panagopoulos (2024) highlighted UA's contribution to fight against food security problems and environmental issues in urban areas. Similarly, Abebe (2024) highlighted UA's potential to contribute to Ethiopia's food security. UA can have potential for localization of food supply as Zhong et al.'s (2023) research emphasized the importance of localization of food supply for urban food security in China. Furthermore, UA is perceived to be critical in future cities' food security as UA has been critical in crisis or conflicts (Barthel et al., 2015 as cited in Pueyo-Ros et al., 2024). For example, UA has contributed to food security in post-earthquake time period (Slater & Birchall, 2022). Additionally, UA is considered to support different aspects of sustainability in cities (e.g., employment opportunities) (Pueyo-Ros et al., 2024).

Effective, sustainable, resilient and smart agriculture in cities can cover many different facilities (e.g., climate resilient storage areas for seeds, hydroponics, aquaponics agriculture, underground agriculture), which can contribute to reduced need for food transportation reducing embodied carbon of agricultural products. For example, Opoku et al. (2024) highlighted green urban space and office spaces' potential for urban farms which can support food self-sufficiency, achievement of UNSDGs 2 and 13 and fight against climate change. Opoku et al. (2024) emphasized urban farms' role in food security and their support for SDGs' achievement through relevant planning regulations/policies. Regarding hydroponic system, Gumisiriza et al.'s (2023) research highlighted hydroponics' role in food security and its potential to perform as conventional farming. Furthermore, regarding vertical farming, Zhang et al. (2023a) focused on skyscraper crop factory as vertical farming system and a sky farm and highlighted its contribution as a crop-production system to reduce GHGs, to increase arable area, to support food security in case of crises and to meet increasing food demand. Furthermore, underground agriculture for extreme adverse situations and risks (e.g., war, climate disaster). Additionally, biomass facilities based on food waste collected can contribute to sustainability through reduced environmental footprint of energy generation, through supported social sustainability (e.g. contributing to reducing fuel poverty) and through economic sustainability (e.g. contributing to energy supply in a sustainable way).

Food security, urbanization and UA relation cannot be considered without effective, efficient and resilient agriculture in rural areas as both urban and rural areas effective and strategic integration can contribute to food security, emphasizing urban and rural areas need to be considered in an integrated way to enhance food security and to overcome food security challenges. Sertyesilisik (2023) highlighted villages' potential for reducing the income gap and supporting social sustainability. Food security's political economy and sustainable agriculture and environment policies need to be considered in an integrated way (Sertyesilisik, 2022). Furthermore, sustainable agriculture and environment policies are within the scope of sustainable development policies (Sertyesilisik, 2022). For this reason, food security policies and strategies are related with development plans and strategies of urban and rural areas covering food security influencing potential in entire food resources.

Recommendations for Enhancing Contribution of the AEC Industry and Its Professionals to the Food Security

Agricultural policies have potential for mitigation of climate change's adverse effects through reduced production costs and supported risk resistance and motivation of farmers (Lee et al., 2024). Countries having higher global food security scores are the ones which enable their farmers to have access to agricultural inputs, and financial products, and the ones which have strong supply chain infrastructure and which have their governments' investment in research and development and innovative technology (Economist website, n.d.). Lee et al. (2024) suggested policies to achieve strengthened climate change resilience, improved system design and coordinated regional agricultural production. Complying with Sertyesilisik (2022) and Sertyesilisik (2023), effective and strategic agricultural policies and country's sustainable development are interrelated in food security performance in entire country at the urban and rural levels.

Agricultural policies need to cover multiple dimensions and aspects related with food security. For example, Meliko et al. (2023) recommended multiple indicators for measuring food security and policies on food insecurity to support production capacity, training and employment opportunities. Focusing on Bangladesh, Ahmed (2024) suggested that raising livestock and agricultural income are complementary and that enhanced education can support households' food security. Furthermore, emphasizing UA's potential for earthquake preparedness, Slater and Birchall (2022) indicated that even if UA is important for improvement in the recovery through improved food security in the post-earthquake phase, UA is not covered in the earthquake/disaster preparedness programmes of many communities (Slater & Birchall, 2022). Additionally, effective agriculture policies and investments need to be integrated with innovation and technology fostering policies and investments. Effective and innovative technology integration's potential for contributing to the food security has been emphasized by many researchers. For example, Rehman et al.'s (2023) research revealed that integration of IoT and blockchain technologies has potential for supporting sustainable practices and food security. Blockchain technology can be used to record and trace/monitor information on crop production enabling easy consumer choices (Rehman et al., 2023). Furthermore, focusing on China's grain production system, Liu and Ren's (2023) research revealed that digital inclusive finance supported food supply's coordinated development level, spatial spillover effect on neighbouring areas and reduction in carbon emissions. Additionally, innovation is vital for building resilience (Economist website). For this reason, sustainable and resilient UA designed and constructed by the effective and expert interdisciplinary team can have potential to enhance to food security in an innovative, strategic and effective way covering all dimensions and aspects of food security.

Agricultural policies need to be designed with the contribution of and considering AEC industry professionals as they are among the main stakeholders of food security. For example, they can contribute to food security through design and construction of effective, sustainable and resilient facilities and infrastructure. Furthermore, they can support food security through supplying necessary construction materials used in and necessary for these facilities and infrastructures. Additionally, they can take part in the establishment of agriculture friendly sustainable, resilient cities, neighbourhoods and villages fostering their potential contribution to and role in food security, as well as in transformation of cities, neighbourhoods and villages into agriculture friendly ones enhancing their food security. They can design and construct agriculture-related work environments considering not only efficiency and effectiveness of the agricultural processes but also motivation, efficiency, safety, effectiveness of the staff and managers working in these places. In this way, they can enhance interior design of these places in human and agriculture friendly ways supporting food security. These places potential to be visited by citizens and villagers under the guidance of experts can motivate their contribution to food security and increase their awareness of the effective and efficient agriculture production techniques and options, and of their potential roles in food security. In this way, these places can further act as training and collaboration areas for these participants in fostering food security. Furthermore, AEC industry professionals, as stakeholders of food security, can be affected by the food security as human beings relying on agricultural outputs' quality and availability.

AEC industry, its professionals and urban planners, and their knowledge and skills on agriculture, and their ability to work in interdisciplinary and effective teams related with all aspects of agriculture can have impact on sustainable and resilient agriculture, its urban and rural infrastructure and facilities vital for food security, and UA and localization of food supply. In this way, they can have the potential for contributing to food system's transformation. Global food system needs to be transformed so that food can be produced sustainably (Liu & Ren, 2023). Agriculture is vital for the global food security (Rehman et al., 2023). Food supply localization is related with connection of urban consumers with producers via short food supply chains and direct marketing (Zhong et al., 2023). From this aspect, UA can have potential for localization of food supply.

Agricultural policies need to cover food security-related education and training of AEC industry professionals, undergraduate and graduate students who are/will be working in the AEC industry and/or taking part in the research and development activities so that their contribution to food security can be enhanced. These education and training programmes need to enable them to be equipped with the skills and knowledge of sustainable and resilient agricultural practices and technologies so that they can effectively integrate these practices and technologies into greenhouses, built environments and agricultural facilities in an efficient, effective and sustainable way. Furthermore, their skills and knowledge related with these practices and technologies can support them to effectively supply construction materials and to effectively design and construct infrastructure, buildings, offices and facilities necessary for food security in the urban and rural areas. As AEC industry professionals and urban planners equipped with skills and knowledge to effectively contribute to the food security are important for food secure cities and villages, their education at the undergraduate and graduate levels and in formal and informal ways throughout their lifetime can support their effective participation in food security and their relevant skills and knowledge to be kept up-to-date.

From the urbanization related perspectives, AEC industry professionals and urban planners need to be equipped with necessary skills and knowledge to support effective, strategic and resilient food secure urban and rural planning considering and assessing urban and rural food security, and integrating UA and food security requirements into urban and rural plans. For example, landscape planning and policy strategies based on food security are vital for urban-rural landscapes' sustainability (Zhang et al., 2023b). Furthermore, Shi et al. (2022) emphasized the contribution of global slope climbing of urban expansion to save cropland. UA has many dimensions/aspects that need to be assessed and optimized by an interdisciplinary stakeholder team including but not limited to the AEC industry professionals and urban planners. For example, in a city in Amazonia covers many aspects (e.g., rational use of urban land, environmental education) (Vieira & Panagopoulos, 2024). As UA's contribution are influenced/affected by many factors (e.g., garden typology, location) (Pueyo-Ros et al., 2024), the interdisciplinary team including the AEC industry professionals and urban planners need to consider and assess these factors effectively.

AEC industry professionals and urban planners equipped with necessary skills and knowledge to support ecological security in rural and urban areas can have potential for supporting food security. Many researchers emphasized importance of ecological security. For example, Zhang et al. (2024) indicated that the determination of ecological security patterns can be effective in ecological and environmental security and that ecological security protection is within the scope of China's development strategies. Furthermore, Zhao et al. (2024) highlighted the importance of scientific ecological security patterns, limited urban expansion boundaries and assessment of ecosystems' health for ecological security.

AEC industry professionals and urban planners need to consider and assess food security supporting technology integration to the rural and urban areas at the buildings and neighbourhood levels in a coherent way. For example, the fourth industrial revolution and smart cities can contribute to solve food insecurity challenge of cities in the future (de Amorim et al., 2019). Similarly, Bin Alam et al. (2023) emphasized that agriculture 4.0 can contribute to food supply and food security. Furthermore, Sharma et al.'s (2024) research on United Arab Emirates revealed digital technology moderated innovative urban farming system-sustainable urban farming relationship.

AEC industry professionals and urban planners equipped with necessary skills and knowledge to effectively contribute to strategic planning and sustainable development plans related with food secure living environments in rural and urban areas can influence their food security levels which can influence countries' and the world's food security. Sertyesilisik (2022) highlighted the importance of interdisciplinary collaboration in food security and relevant sustainable development policies. Importance of strategic planning and development plans in food security has been revealed in many researches (e.g., Lee et al., 2024; Pueyo-Ros et al., 2024). Strategic planning can be used for enhancing UA's support for urban sustainability (Pueyo-Ros et al., 2024). Strategies for urban food security need to focus on systematic and integrated approaches to support social security programmes (Dinku et al., 2023). For all these reasons, AEC industry professionals and urban planners' contribution to strategic planning, sustainable development plans and agricultural policies to enhance food security of their countries can have potential to comply with their countries' effectiveness and success in enhancing food security. This potential synergy can be useful in achieving UNSDGs related with food security influencing sustainability. This situation emphasizes relationship among and interaction of food security, agricultural policies, sustainable development plans and expert AEC industry professionals in supporting food security and in achieving enhanced sustainability and resilience of humanity.

Effective and strategic education and training of students, researchers, academics and professionals of AEC industry on agriculture integrated built environment and relevant technologies can support establishment of agriculture integrated built environment in cities and villages and transformation of built environment into agriculture integrated ones (Fig. 14.1) which can further support and contribute to food security (Fig. 14.1). The undergraduate curriculum can cover and be enriched by elective and/or compulsory courses on requirements of built environment to support local agriculture; agriculture integrated and enabling built environment in cities and villages; technology needed to enhance agriculture in the built environment. Furthermore, interdisciplinary master degree programmes specific for enhancing postgraduate students' expertise on effective and strategic integration of agriculture and relevant contemporary technologies into built environment at different levels (e.g., building, neighbourhood, city and village levels). Interior architects (e.g., interior design of offices of UA facilities and interior design of agriculture integrated built environment), architects (e.g., design of greenhouses and UA facilities), engineers (e.g., civil engineers) can contribute to infrastructure, and built structures needed for renewable energy based agriculture integrated built environment. Additionally, relevant life-time continuous trainings of AEC industry professionals in formal and informal education with respect to agriculture-related technologies' integration to the built environment can have potential to support them to gain knowledge and skills related with agriculture integrated built environment and to be kept-updated with changes in this field so that they can effectively contribute to widespread of agriculture in cities and villages. These education and training programmes can be face-to-face and/or they can be supported by online education and training programmes as well. Virtual and online education and training can support widespread of these education and training programmes to reach to target groups effectively and to get benefit from and to create synergy from international and interdisciplinary experience and expertise (Sertyesilisik & Sertyesilisik, 2023). Furthermore,

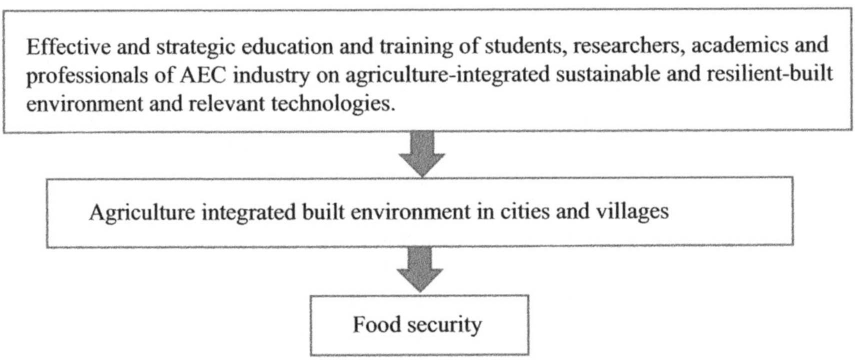

Fig. 14.1. Relationship among Agriculture-Integrated Built Environment, Effective and Strategic Education of AEC Industry Professionals, and Food Security. *Source*: Sketched by the author.

widespread of these effective and strategic education and training programmes beyond the AEC and agriculture industry and professionals and academics to all stakeholders of food security and of the renewable energy-based agriculture integrated sustainable and resilient-built environment can have potential to transform demand and supply in a way to encourage the widespread of this type-built environment.

Expert AEC industry professionals and researchers can have potential to effectively enhance efficiency and effectiveness of their designs and constructions' agriculture output performance through assessing-built environment specific local aspects (e.g., climatic condition, local vegetables and fruits, flowers) and other technologies protecting agricultural production process and outputs from adverse climatic conditions local aspects. In this way, their conscious design decisions can have potential to influence efficiency and effectiveness of agricultural production contributing to food security.

Expert AEC industry and agriculture professionals and academics can have strategic contribution to effective and efficient renewable energy-based agriculture integrated sustainable and resilient-built environment fostering sustainability, sustainable and resilient development, and sustainability and resilience enhanced built environment (Fig. 14.2). Renewable energy-based agriculture-integrated sustainable and resilient-built environment can act as a multiplier factor for sustainable and resilient development (Fig. 14.2). Furthermore, renewable energy-based agriculture-integrated sustainable and resilient-built environment can act as a way to support all pillars of sustainability. From the social sustainability perspective, for example, renewable energy-based agriculture-integrated sustainable and resilient-built environment can have potential to contribute to health and well-being of users of this built environment. They can benefit from the healthy foods they produce. They can get benefit from the potential of increase in relative self-sufficiency and potential of reduction in food wasted. They can feel themselves to be connected with nature. Furthermore, from the economic sustainability perspective, for example, this type of sustainable- and resilient-built environment can

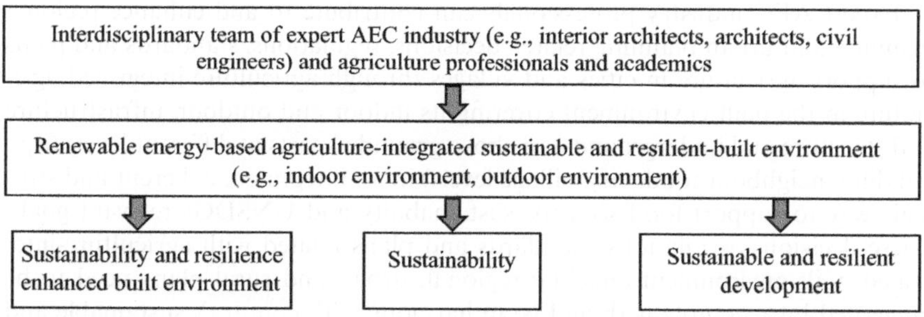

Fig. 14.2. Potential of Expert AEC Industry and Agriculture Professionals and Academics to Act as Drivers for Sustainability, Sustainable and Resilient Development, Sustainability and Resilience-Enhanced Built Environment. *Source*: Sketched by the author.

have potential to enhance agricultural production and to create job opportunities related with agriculture activities integrated in the built environment. Additionally, from the environmental sustainability perspective, it can have potential to contribute to reduction in embodied carbon due to supported local agriculture supply and demand. These examples for renewable energy-based agriculture-integrated sustainable and resilient-built environment's potential contribution to all pillars of sustainability reveal its potential to support and be integrated part of sustainable and resilient development plans at the urban and rural levels and at the country level.

Expert interior architects can support renewable energy-based agriculture integrated sustainable and resilient healthy indoor environment. For example, in houses and offices, this type of micro-scale agriculture-integrated spaces can have potential to enhance indoor air quality, keep occupants to connect with nature, feel connected and stay connected with nature potentially supporting health and well-being and reduction in stress. Furthermore, interior architects can enhance their interior design via neurodesign in indoor environment of agriculture-integrated interior environment in a way to enhance health and well-being of occupants. Additionally, these indoor environments can be designed in a way to enable occupants to experience agriculture at micro scale. These spaces, if designed effectively and considering neurodesign, they can have potential to increase efficiency, and motivation in the work environment as they can enable staff to collaborate in micro-scale indoor agriculture activities and as they can provide the staff an environment to regain their energy. For example, in offices, effectively designed café break rooms where agriculture is integrated can be effective in motivation of staff as well.

Expert architects can support establishment of renewable energy-based agriculture-integrated sustainable and resilient-built environment (e.g., building, neighbourhood, city and urban) and transformation of built environment into renewable energy-based agriculture-integrated sustainable and resilient-built environment one. Expert architects need to assess local factors (e.g., local climatic conditions, microclimate) specific to built environment where agriculture will be integrated to enhance efficiency of agriculture activities in the built environment and to enhance indoor air quality of the built environment relying mainly on passive design.

Expert AEC industry professionals can contribute to and enhance regional planning and urban planning related decisions, regulations, standards and plans to support agriculture in cities and villages through agriculture integrated agriculture in the built environment covering its indoor and outdoor, infrastructure and structure and underground and above ground aspects at different levels (e.g., building, neighbourhood, city, village levels) in an integrated, coherent and strategic way to support food security, sustainability and UNSDGs relevant goals. These decisions, regulations, standards and plans related with agriculture-integrated built environment enabling regional, urban and rural plans need to be integrated into, comply with and be in harmony with country's sustainable and resilient development plans covering many different interacting aspects (e.g., education, investment, industrial development, research and development aspects) strategically to contribute to food security efficiently and effectively. In this way, expert AEC industry professionals and academics can support regulations related with renewable energy-based agriculture-integrated built environment covering different levels (e.g., building, neighbourhood, city, village levels) and unique aspects of each built environment's specific location related factors (e.g., climatic conditions) in a coherent and integrated way.

Conclusion

This chapter emphasized AEC industry's roles in, contribution to and vitality for food security, sustainable and resilient agricultural practices. This chapter emphasized relationship among agriculture-integrated built environment, effective and strategic education of AEC industry professionals, and food security (Fig. 14.1). This chapter underlined importance of and need for undergraduate and graduate curriculums, interdisciplinary master degree programmes and formal and informal learning programmes for future AEC industry professionals to support their skills and knowledge of sustainable and resilient agricultural practices and technologies-integrated greenhouses and built environment so that they can design and construct relevant spaces (e.g., indoor and outdoor) and their infrastructure in human and agriculture friendly way supporting food security in cities and rural areas, and people's awareness of and contribution to food security as these places can act as agriculture producers and training and collaboration areas as well. Furthermore, they can contribute to transformation of cities and villages to become more effective in food security and to establish new ones which are food secure. These expert AEC industry professionals, as among the main stakeholders of food security, and sustainable and resilient UA, can have the potential for enhancing food security and resilience against and minimization in climate change caused and other challenges affecting achievement of UN SDGs. For this reason, AEC industry and its professionals' ability to support food security in compliance with United Nations SDGs and countries' development plans is one of the key factors that can influence all pillars of sustainability, well-being and welfare, sustainability and resilience of the people, and sustainable development. This chapter emphasized potential of expert AEC industry and agriculture professionals and academics to act as drivers for sustainability, sustainable and resilient development, sustainability and resilience enhanced built environment (Fig. 14.2). They can contribute to fulfilment of countries' food security capacity. Furthermore, relationship among and interaction of food security, agricultural policies, sustainable development plans and AEC industry professionals in supporting food security and in achieving enhanced sustainability and resilience of humanity can have potential to get benefit from expert AEC industry professionals' effective contribution to agriculture policies, regulations and standards (e.g., related with renewable energy based agriculture integrated built environment) supporting sustainable and resilient development. For this reason, agricultural policies need to cover food security-related interdisciplinary education and training of AEC industry professionals. Furthermore, agricultural policies need to be designed with the contribution of and considering AEC industry professionals as they are among the main stakeholders of food security. This chapter can be useful for all stakeholders of the food security supported by sustainable and resilient agricultural practices and policies covering sustainable development plans and strategies.

References

Abebe, M.G. (2024). Impacts of urbanization on food security in Ethiopia. A review with empirical evidence. *Journal of Agriculture and Food Research*, 15, 100997.

Ahmed, R. (2024). Climate shocks' impact on agricultural income and household food security in Bangladesh: An implication of the food insecurity experience scale. *Heliyon*, https://doi.org/10.1016/j.heliyon.2024.e25687

Armanda, D. T., Guinée, J. B., & Tukker, A. (2019). The second green revolution: Innovative urban agriculture's contribution to food security and sustainability – A review. *Global Food Security*, *22*, 13–24.

Badami, M. G., & Ramankutty, N. (2015). Urban agriculture and food security: A critique based on an assessment of urban land constraints. *Global Food Security*, *4*, 8–15.

Barthel, S., Parker, J., & Ernstson, H. (2015). Food and green space in cities: A resilience lens on gardens and urban environmental movements. *Urban Studies*, *52*(7), 1321–1338. https://doi.org/10.1177/0042098012472744

Bin Alam, F., Zaman, S. R., Gonzalez, E. D. R. S., Bari, A. B. M. M., & Karmaker, C. L. (2023). Analysis of the drivers of Agriculture 4.0 implementation in the emerging economies: Implications towards sustainability and food security. *Green Technologies and Sustainability*, *1*, 100021.

de Amorim, W. S., Deggau, A. B., do Livramento Gonçalves, G., da Silva Neiva, S., Prasath, A. R., & de Andrade Guerra, J. B. S. O. (2019). Urban challenges and opportunities to promote sustainable food security through smart cities and the 4th industrial revolution. *Land Use Policy*, *87*, 104065.

Dewan, S., Bamola, S., & Lakhani, A. (2024). Addressing ozone pollution to promote United Nations sustainable development goal 2: Ensuring global food security. *Chemosphere*, *347*, 140693.

Dinku, A. M., Mekonnen, T. C., & Adilu, G. S. (2023). Urban food systems: Factors associated with food insecurity in the urban settings evidence from Dessie and Combolcha cities, north-central Ethiopia. *Heliyon*, *9*, e14482.

Filippini, R., Mazzocchi, C., & Corsi, S. (2019). The contribution of Urban Food Policies toward food security in developing and developed countries: A network analysis approach. *Sustainable Cities and Society*, *47*, 101506.

Goal 2: End hunger, achieve food security and improved nutrition and promote sustainable agriculture. https://unstats.un.org/sdgs/report/2016/goal-02/

Gumisiriza, M. S., Ndakidemi, P. A., Nampijja, Z., & Mbega, E. R. (2023). Soilless urban gardening as a post covid-19 food security salvage technology: A study on the physiognomic response of lettuce to hydroponics in Uganda. *Scientific African*, *20*, e01643.

Global Food Security Index (2022). https://impact.economist.com/sustainability/project/food-security-index/resources/Economist_Impact_GFSI_2022_Website_Navigation_Guide_Sep_2022.pdf

Economist (n.d.). Key Findings https://impact.economist.com/sustainability/project/food-security-index#introduction Economics Impact

Lee, C.C., Zeng, M., & Luo, K. (2024). How does climate change affect food security? Evidence from China. *Environmental Impact Assessment Review*, *104*, 107324.

Liu, J., & Ren, Y. (2023). Can digital inclusive finance ensure food security while achieving low-carbon transformation in agricultural development? Evidence from China. *Journal of Cleaner Production*, *418*, 138016.

Li, S., Ji, Q., Liang, W., Fu, B., Lü, Y., Yan, J., Jin, Z., Wang, Z., & Li, Y. (2023). Urbanization does not endanger food security: Evidence from China's Loess Plateau. *Science of the Total Environment, 871*, 162053.

Meliko, M. O., Mossy, E. R., & Ngaiwi, M. E. (2023). Food accessibility measurements amongst rural and urban informal dwellers in Buea Municipality. *Journal of Agriculture and Food Research, 12*, 100606.

Mkhize, X., Mthembu, B. E., & Napier, C. (2023). Transforming a local food system to address food and nutrition insecurity in an urban informal settlement area: A study in Umlazi Township in Durban, South Africa. *Journal of Agriculture and Food Research, 12*, 100565.

Opoku, A., Duff, A., Yahia, M. W., & Ekung, S. (2024). Utilisation of green urban space for food sufficiency and the realisation of the sustainable development goals – UK stakeholders perspective. *Geography and Sustainability, 5*, 13–18.

Pueyo-Ros, J., Skerjanec, M., Castellar, J. A. C., Atanasova, N., Comas, J., & Corominas, L. (2024). Beyond food: A stochastic model to estimate the contributions of urban agriculture to sustainability. *Landscape and Urban Planning, 241*, 104930.

Rehman, K. U., Andleeb, S., Ashfaq, M., Akram, N., & Akram, M. W. (2023). Blockchain-enabled smart agriculture: Enhancing data-driven decision making and ensuring food security. *Journal of Cleaner Production, 427*, 138900.

Rezai, G., Shamsudin, M. N., & Mohamed, Z. (2016). Urban agriculture: A way forward to food and nutrition security in Malaysia. Urban planning and architecture design for sustainable development, UPADSD 14–16 October 2015. *Procedia Social and Behavioral Sciences, 216*, 39–45.

Sertyesilisik, E. (2022). Gıda Güvencesini Arttırmaya Yönelik Sürdürülebilir Tarım ve Çevre Politikaları. *Süleyman Demirel Üniversitesi Vizyoner Dergisi, 13*(36), 1394–1406.

Sertyesilisik, B. (2023). Sustainable and smart villages' roles in reducing the income gap and supporting socio-economic indicators. In R. C. Das (Ed.), *Inclusive developments through socio-economic indicators: New theoretical and empirical insights* (pp. 193–202). Emerald Publishing Limited. https://doi.org/10.1108/978-1-80455-554-520231015

Sertyesilisik B., & Sertyesilisik E. (2023). Political economy of the 'virtual' internationalisation of higher education to foster glocal sustainable development. *Journal of Learning and Teaching in Digital Age, 8*(1), 1–9.

Sharma, R., Wahbeh, S., Sundarakani, B., Manikas, I., & Pachayappan, M. (2024). Enhancing domestic food supply in the UAE: A framework for technology-driven urban farming systems. *Journal of Cleaner Production, 434*, 139823.

Shi, K., Wu, Y., & Liu, S. (2022). Slope climbing of urban expansion worldwide: Spatiotemporal characteristics, driving factors and implications for food security. *Journal of Environmental Management, 324*, 116337.

Slater, T., & Birchall, S.J. (2022). Growing resilient: The potential of urban agriculture for increasing food security and improving earthquake recovery. *Cities, 131*, 103930.

Vieira, T. A., & Panagopoulos, T. (2024). Urban agriculture in Brazil: Possibilities and challenges for Santarem, eastern Amazonia. *Land Use Policy, 139*, 107082.

Zhang, L., Huang, L., Li, T., Wang, T., Yang, X., & Yang, Q. (2023a). The skyscraper crop factory: A potential crop-production system to meet rising urban food demand. *Engineering*, 31, 70–75.

Zhang, Y., Li, J., Liu, X., Bai, J., & Wang, G. (2023b). Do carbon sequestration and food security in urban and rural landscapes differ in patterns, relationships, and responses? *Applied Geography, 160*, 103100.

Zhang, B., Cao, J., Chen, D., Li, X., Liu, Y., Wang, J., & Liu, T. (2024). Construction of watershed ecological security patterns with integrated of spatial variability: A case study of the Yellow River Basin, China. *Ecological Indicators, 159*, 111663.

Zhao, Y., He, L., Bai, W., He, Z., Luo, F., & Wang, Z. (2024). Prediction of ecological security patterns based on urban expansion: A case study of Chengdu. *Ecological Indicators*, *158*, 111467.

Zhong, T., Crush, J., Si, Z., & Scott, S. (2023). The Nanjing model: Comprehensive food system governance, localization and urban food security in China. *Global Food Security*, *38*, 100709.

Chapter 15

Technical Efficiency, Small Pond Fishery and Sustainability: Evidence from a Stochastic Frontier Regression Approach with Reference to West Bengal, India

Kishan Agarwalla[a] *and Tonmoy Chatterjee*[b]

[a]*Department of Economics, University of North Bengal, India*
[b]*Department of Economics, Bhairab Ganguly College, India*

Abstract

The study is undertaken to examine the technical efficiency (TE) of small pond fishery using the stochastic frontier model in the northern region of the state of West Bengal in India. The 65 samples were collected through field surveys in Uttar Dinajpur and Dakshin Dinajpur districts for three months (i.e. from May to June) in the year 2022. The stochastic frontier model estimation indicates that boosting investment in labour, organic fertilizers, fish fingerlings and land area has the potential to enhance returns in fish production. The findings demonstrate that TE spans between 83 and 100 per cent, averaging at 94 per cent. This suggests that, on average, fish farmers in the examined region are operating slightly below the highest achievable production level, falling short by approximately 6 per cent, which raises concerns about sustainability. The study recommends that the use of labour and organic fertilizers (i.e. cow dung) should be made available to transform traditional rearing practices into more productive scientific methods. Also, the land area should be extended and more fish fingerling should be used to increase the production of fish in the study area.

Keywords: Small pond fishery; sustainability; maximum likelihood; frontier production function; technical efficiency; stochastic frontier

Sustainable Agricultural Practices: Economic and Environmental Implications, 223–236

doi:10.1108/978-1-83608-336-820241015

Introduction

One important line of empirical research that has proceeded in the result of the publication of Farrel's work is the stochastic frontier regression (SFR). This approach tries to operationalize Farrell's conceptual framework by specifying the frontier as a parametric function of inputs. It was initially applied for the measurement of technical efficiency (TE). Later on, it has been extended to include allocative, cost and profit efficiency. The first statistical formulation of this method was autonomously proposed by Aigner et al. (1977) and Meeusen and van den Broeck (1977). The specification of the SFR model involves a production function that has a two-component composite error term – one accounts for random effects and another accounting for technical inefficiency. The advantage of this method is that it can make a distinction between technical inefficiency and statistical noise effects that originate from the data set. The stochastic frontier approach is considered very appropriate for assessing TE in a developing country, where data are often heavily influenced by measurement errors and other stochastic factors such as weather conditions, diseases, etc. (Coelli et al., 1998; Fare et al., 1985; Kirkley et al., 1995).

The estimation of the SFR model involves (i) estimating the parameters of the frontier function (x) which is to be specified a priori and (ii) estimating inefficiency. Kumbhakar and Wang (2010) have given a detailed account of the method of estimation of the stochastic frontier production function. They have demonstrated that the method to be applied is the maximum likelihood (ML). An essential requirement for this method is the specification of distributional assumptions of the error terms. Since the limited theory is available in guiding the choice of distributions, distributional assumptions are often made heuristically. The various alternative distribution assumptions explored in the literature so far are an exponential distribution (Meeusen & Van den Broeck, 1977); a normal distribution truncated at zero (Aigner et al., 1977); a half-normal distribution truncated at zero (Jondrow et al., 1982); and a two-parameter gamma/normal distribution (Greene, 1990; Pascoe et al., 2003). It is to be noted that the estimates of TE are dependent on the specific distributional assumption of the error components.

The analysis of TE in the small-scale fishery is relevant for several reasons. Enhancing TE on a farm is crucial for augmenting both production and productivity. While this type of fishery is highly common in developing countries, there is a very limited number of analyses assessing their efficiency. Indeed, there is no precise information on the contribution of small-scale fisheries to livelihoods and economies in developing nations (Worm et al., 2009). Exploring this gap in the research is relevant for the decision-making policy. On the one hand, small-scale fisheries can generate significant profits and be more resilient to shocks and crises – there are mainly two important elements, i.e., poverty alleviation and food security. Fishery management should be linked with efficiency considerations (Grafton et al., 2006). The question is what factors constrain efficiency?

In West Bengal, fish farming is practiced by economically disadvantaged small and marginal farmers who rely on locally accessible resources, resulting in low productivity. It is imperative to evaluate the effectiveness of different inputs employed in fish production. Given this context, this study aims to analyse the efficiency of resource utilization in fish production, farm-specific TE and factors contributing to technical inefficiency in fish farming. The findings will offer valuable insights to formulate policies aimed at optimizing scarce farm resources to enhance both production and profitability of fish farms in West Bengal.

Review of Literature

In the context of North Bengal, the availability of literature is very scanty and therefore we need to include the study related to West Bengal, India and some related studies of the world. However, there is an emerging need for conducting a sample study in order to investigate the various operational aspects of Pond fishery in North Bengal. This can help understand the economic rationality behind emerging smallholders as well as the impacts of such changes on the regional economy of North Bengal. It is also to be mentioned that the examination of efficiency is a central issue in judging the performance of a production unit such as a pond fishery. The studies of efficiency/inefficiency also help us to identify the factors causing inefficiency. But all most no such empirical research has been carried out yet. The proposed study aims to bridge these research gaps.

Roy (1999) analysed the existing marketing of fish fingerlings with special reference to environmental concerns and awareness of the fishermen. The constraints have been identified and suggestions have been made for the development of marketing of fishery on the basis of primary data collected from Dakshin Dinajpur district of West Bengal in 2006. The small and marginal farmers dominate the fishery and about 10 per cent of fish farmers earn their major livelihood from fishing. The share of fishermen has been found to be 62.80 per cent which is higher than any other agricultural crop. The average elasticity has been found to be one (approx.). The fish market was found to be competitive and efficient in the study area. The pollutants in aquatic environments are diverse and complex in nature. The main sources of pollutants are pesticides and leaching of dissolved chemical fertilizers, sewage and water disposal, retting of fibre crops, processing wastes, etc. The degree of consciousness about the caught fish has a declining trend as found in the study area. The study has found that over-fishing and lack of production culture are the main reasons for this decline. To upgrade the socio-economic status of the fishermen in this district, the study suggested collective effort and sustainable fish conservation, production and marketing with the existing resources. Again, Saha (2016) in her study made an attempt to investigate the Economics of Fishery and Livelihood in the Coastal belt of West Bengal. For this, she has emphasized the issue of sustainability of the fisheries sector in coastal belt of West Bengal. This issue has gained importance because of the fact that the fishery management practices in India have experienced different kinds of obstacles in different periods of time in various parts of the country. This issue has also strong implications for the livelihood of the stakeholders associated with fisheries.

Alam et al. (2005) have used translog stochastic frontier cost function to determine the economic efficiency of fish production in Bangladesh. The economic inefficiency effect model was also estimated simultaneously with the stochastic frontier to identify factors that influence efficiency. The coefficients of age and experience were negative and significant in the stochastic frontier which meant that the cost of fish production decreased with the increase in age and experience of the farm operators. The sign of farm size was positive and significant in the inefficiency effect model, which meant that the large farmer was economically less efficient than a small farmer. The economic efficiency varied from 15% to 99% at an aggregate level, 96% to 98% in Mymensingh and 63% to 99% in the Rangpur region. The mean economic efficiencies were 97%, 89%, and 71% for Mymensingh, Rangpur, and all regions, respectively. There appeared to be 3%, 11% and 29% economic inefficiencies for Mymensingh, Rangpur and all regions, respectively. This study indicated that the cost of production could be reduced by 29% keeping the output constant at the aggregated level. For policy options, extension services should be made available for all fish farmers. Education should be another one of the top priorities of the government to develop the necessary human capital for sustainable development of the fishery sector. While Kundu (2011) has pointed out that inland fisheries play a crucial role in the development of the rural economy in west Bengal like any other economic activity. Being endowed with various water resources in the Kumarganj block of Dakshin Dinajpur district, West Bengal has immense potential for the development of inland fishing. This chapter is an attempt to highlight the growth of the fishery of the concerned area in terms of fish production and employment generation based on the data and information from secondary sources. The author has tried to identify the major problems related to inland fishing activities and suggested effective measures for the development of inland fishery in the Kumarganj block of Dakshin Dinajpur district. On the other hand, Olaniyi (2012) has examined the resource-use efficiency of fish production in Oyo State, Nigeria. Multi-stage sampling techniques were used from 100 respondents. Descriptive analysis, gross margin analysis, as well as inferential statistics like stochastic production function, was employed in the analysis of the inefficient level of farmers. The result showed that the gross margin of the average fish farmer was estimated at 20, 4279.72. The stochastic frontier production function analysis showed that coefficient labour (0.1897), numbers of fingerlings (1.0160) and production period (−0.6663) were statistically significant at 5 per cent, 1 per cent and 10 per cent, respectively. The majority of the farmers, which is about 77 per cent, have their efficiencies between 0.5 and 0.9 while the mean efficiency of the fish farmers in the study area was 0.67. In an interesting study, Misra and Misra (2014) have pointed out the concept of meta-frontier data envelopment analysis (DEA) to examine whether there is any systematic difference in the TE of fish farms of diverse size classes categorized on various socio-economic situations. Efforts have also been made to identify the influence of those characteristics on TE scores using regression analysis. The study has shown that the overall mean TE is 62.8 per cent, which indicates that on average, the realized fish output can be raised by 37 per cent in the state with the existing technology and resources. Ample variations in TE scores have been found when farms were categorized on the basis of size, region, ownership pattern or proprietorship. The ownership, farm experience and sole proprietorship are the most important determinants of TE. However, pond size and education have not depicted any major relationship with TE. In order to improve the efficiency of fish culture, location-specific should be adopted in development strategies, long-term leasing policies and participatory extension should support West Bengal.

Hossain et al. (2015) have emphasized that agricultural productivity varies due to several reasons – differences in production technology, differences in the setting in which production occurs and differences in the efficiency of the production process. The efficiency measurement has been the concern of researchers with an aim to investigate the efficiency levels of farmers engaged in agricultural activities. The major task is to identify the efficiency determinants in the efficiency analysis. The main objective is to estimate the efficiency of the small banana grower in the Kushtia district of Bangladesh. In this study, econometric frontier models were used under the specification of the Cobb-Doulas production function model. Primary data were collected by the use of a well-structured questionnaire. Among all elasticities, the elasticity for land used for banana production is the largest (0.3498). The second largest elasticity is the cost of fertilizer which is 0.2417 and the elasticity for the cost of labour is the lowest (0.0495). It is revealed from the study that the age and education level of the farmers has a positive effect on TE. The government also should take the necessary steps to improve the knowledge of a farmer about modern cultivation techniques and encourage them to adopt new technology to increase production. In this context, Ekramuddin (2017) has conducted a survey in 10 significant villages of Birbhum, an economically backward district of West Bengal to analyse the impact of small-scale fish farming in the rural socio-economic aspect. Five respondents from each village were interviewed providing a set of questionnaires to obtain a clear view of demography, social and financial status, the nature of the fishery, and the economy that is directly related to the fishery. It reveals that the majority of fish farmers are males and most are not educated beyond class 10 levels. The respondents have an average household size ranging between three and six persons. The result of descriptive analysis and test of significance showed that there is a significant impact of small-scale fishery enterprise on the rural economy. In a recent study, Majumder (2017) examined the economics of tea growing on small land holdings in West Bengal. The major thrust area of the study has been taken to be a parametric as well as nonparametric analysis of TE in order to make an objective assessment of the productivity performance of this emerging sector. For the parametric analysis of efficiency, the stochastic production frontier approach has been employed. For the non-parametric study of efficiency, on the other hand, the deterministic DEA technique has been used. The results of the analyses reveal that the overall efficiency level of small tea-growing operations is quite satisfactory. This finding is further substantiated by the decomposition analysis of overall efficiency. The zone-wise analysis of mean TE scores also corroborates this result. The major economic implication to be strained from efficiency analysis is that tea growing on small landholdings has emerged as an efficient model of commercial farming replacing traditional agricultural farming in a phased manner in different areas of the region. Similarly, Moges et al. (2018) determined the level of TE of smallholder teff producers and identify factors affecting the TE of smallholder farmers in teff production of Jamma district, South Wollo Zone, Ethiopia. A three-stage sampling technique was used to select 149 sample farmers. A Cobb–Douglas stochastic frontier production analysis approach was used to estimate TE and to identify the determinants of the efficiency of teff-producing farmers. The ML parameter showed that teff output was positively and significantly influenced by area, fertilizer, labour and number of oxen. The mean levels of TE of the sample farmers were 78%. So, there exists a possibility to increase the level of teff output by 22% through efficiently utilizing the existing resources. The estimated stochastic production frontier model together with the inefficiency parameters showed that age, education, improved seed, training and credit were found to have negative and significant effects on technical inefficiency while farm size was found to have a positive and significant effect on technical inefficiency of teff production. Hence, local government should provide necessary support such as formal as well as informal education, training, credit, improved seed and timely supply of fertilizer. In a more recent study, Singh and Devi (2020) investigated the production function and TE using a stochastic frontier model for carp production in Manipur. They collected primary data from 150 carp producers across three districts of Manipur during the period 2016–2017. The analysis of the production function revealed that increasing investment in fish fingerlings and feed could enhance returns from carp production. Factors such as experience in carp production, level of education and access to credit significantly influenced TE. The findings indicated that TE varied between 21 and 98 per cent, with an average efficiency of 72 per cent. This suggests that the average fish farmer in the study area is operating below the maximum potential production level by approximately 28 per cent.

In summary, these studies collectively highlight the economic significance of fisheries, the factors influencing production efficiency and the need for sustainable development strategies in regions like North Bengal. The proposed study aims to contribute to bridging existing research gaps by examining operational aspects and efficiency in pond fishery, thereby informing policy and development initiatives.

Materials and Methods

The study is based on primary data. The aim of this study was to estimate the TE of selected fishermen in the northern region of West Bengal. The 65 samples were collected through a field survey in the region of Uttar Dinajpur and Dakshin Dinajpur for three months (i.e. from May to June) in the year 2022 in this region. These two districts were selected from all seven districts of North Bengal's region. The selections were based on the huge production in these two districts. In the second stage, each block was selected from each district and then two villages from each block were purposively chosen. At the final stage, a comprehensive list of pond or tank owners from each village in that block was prepared. In case of non-fulfilment of this condition in a particular village, a cluster of villages with two or more contiguous villages was constituted. From the first block, a total of 35 samples were collected randomly (i.e., 20 samples from the first village and 15 samples from the second village. From the second block, 30 samples were collected randomly of which 15 samples were collected from each village. In this way, a total of 65 pond-owner households were selected from these two districts.

The study was conducted on the basis of an open-ended and exploratory research design and data were collected using pre-designed structured questionnaires.

Theoretical Framework

The stochastic frontier production function, commonly referred to as the 'Composite error model', was utilized to assess the fish farm's specific TE. Aigner et al. (1977) and Meeusen and Van Den Broeck (1977) introduced the stochastic frontier production function, which incorporates two disturbance terms: one to address technical inefficiency and the other to accommodate random events impacting production. It is formulated as follows:

$$Y_i = f(X_i;\beta)\exp(V_i - U_i) \tag{1}$$

where Y_i describes output of the *i*th farm; X_i denotes the vector of quantities of each inputs for the *i*th farm; β stands for to represent the vector of unknown parameters prevails in the above-stated set up; V_i indicates the symmetric component of the random error term for the *i*th farm, while U_i represents the random variable with positive sign under the control of the representative farm.

Utilizing the density functions of U_i and V_i, the frontier production function can be estimated via the ML method. The error term $V \sim N(\sigma^2 v)$ represents a symmetrical two-sided error term, with effects distributed across ($-\infty < V_i < \infty$), signifying random shocks beyond the farmer's control. On the other hand, the independent error component U_i is presumed to be non-negative, denoting TE. This error term is one-sided, truncating the distribution $N(0, \sigma^2 u)$. Economically, U_i signifies that each farm's production must align either on or below the production frontier. The disparity between a farmer's production level and the estimated production level dictated by the frontier production function is evaluated using the variance ratio (γ), expressed as $\gamma = \sigma^2 u / \sigma^2 v$. This ratio ranges from zero to one, where $\sigma = 0$ indicates the absence of inefficiency, and $\sigma = 1$ implies the absence of random noise.

Empirical Model

Below is the stochastic frontier production function of the Cobb–Douglas type utilized in the study:

$$\ln Y = \beta_0 + \beta_1 \ln X_1 + \beta_2 \ln X_2 + \beta_3 \ln X_3 + \beta_4 \ln X_4 + V_i - U_i \tag{2}$$

where

Y is the total fish production (kg s^{-1} Bigha), X_1 the total fish fingerlings (kg s^{-1} Bigha), X_2 the total land's area (in Bigha), X_3 the total organic fertilizer (in kgs), X_4 the labour units (labour hour), U_i the farm-specific technical inefficiency effect in the production process,

V_i the random variations due to factors beyond the control of the fish farmer,

β_0 the constant parameter, β_1, β_2, β_3 and β_4 the coefficients to be estimated, ln the natural logarithm and i refers to the ith farm in the sample.

The equation for the technical inefficiency model (U_i) is given below

$$U_i = \delta_0 + \delta_1 \ln W_1 + \delta_2 \ln W_2 + \delta_3 \ln W_3 \tag{3}$$

where U_i is the technical inefficiency of the ith fish farm, W_1 the age of the fish farmer (in years); W_2 the experience of the fish farmer (in years) and W_3 the family size (in no.).The intercept (δ_0) and the coefficients (δ_1, δ_2 and δ_3) of the independent variables are parameters requiring estimation. Maximum likelihood estimates (MLE) of these parameters for both the stochastic frontier production function and the inefficiency model were obtained simultaneously (Coelli et al., 1998). The parameters underwent significance testing at levels of 1 per cent, 5 per cent and 10 per cent.

Empirical Results and Analysis

Socio-Economic Features of Fish Farmers

The respondents' average age was 43 years, indicating that fish farming is primarily undertaken by middle-aged and older individuals, with relatively few young people engaging in it. This aligns with findings by Asamoah et al. (2012). On average, each household consisted of four persons, with family members often contributing as labour for the upkeep and management of fish farms. A notable trend was the low educational attainment among farmers, with 41 per cent having education below matriculate level and only 24 per cent attaining education beyond metric. This educational background seemed to influence the adoption of new knowledge and skills, which appeared lacking in the region.

The average pond size was 1.23 hectares, and farmers typically had 11 years of experience in the profession. For 85 per cent of the households surveyed, fishery constituted the primary occupation, highlighting aquaculture as the mainstay livelihood for most farmers. The adoption of composite fish culture, combining indigenous and exotic carp with diverse feeding habits, aimed at maximizing yields compared to monoculture. Among the various fish species reared, rohu was the most preferred by 90 per cent of farmers, followed by mirgal (87 per cent) and grass carp (70 per cent). Additionally, common carp, silver carp and tilapia were commonly raised species. Similar preferences for composite carp culture over monoculture were also noted in studies by Umamaheswari et al. (2017) and Singh (2015).

Production Analysis

The core idea behind frontier production function analysis lies in pinpointing the farm that utilizes inputs most efficiently, with other farms then measured against this benchmark to determine their efficiency relative to the best-performing one. TE, in this context, signifies the degree to which farms effectively transform limited inputs into tangible output. The outcomes of the stochastic frontier production function estimates are detailed in Table 15.1.

Table 15.1. Maximum Likelihood Estimates of the Stochastic Production Function for Fish Production.

Variables and Parameters	**ML Estimate**		
	Co-efficient	**Standard Error**	***t*-Ratio**
Constant	−0.0048	0.0368	−0.1303
Fish fingerlings	0.2346	0.1034	2.2692**
Land's area	0.3248	0.1728	1.8796***
Organic fertilizer	0.5037	0.1881	2.6777*
Labour	18.58	1.0346	17.9613*
Inefficiency model			
Constant	0.3127	0.1371	2.2808
Age	0.0001	0.0027	0.0572
Experience	−0.0121	0.0072	−1.6666
Family size	−0.0325	0.0197	−0.0164
Variance parameters			
Sigma-squared	0.0266	0.0048	5.502
Gamma	0.0039	0.0051	0.7544
Log-likelihood function	25.5836		
LR test of the one side error	6.4876		

Source: Self-computation from field survey, 2022.
Note: *** denotes significance at 1 per cent, ** denotes significance at 5% and * denotes significance at 10%.

The calculated gamma (γ) parameter was found to be statistically insignificant, indicating the absence of technical inefficiency in fish production. This suggests that the majority of the observed output deviation from the frontier was not due to random stochastic factors. With an estimated value of $\gamma = 0.0039$, it appears that only a minimal portion of the output variation among fish farmers is attributable to technical inefficiency, leaving approximately 99 per cent as influenced by other external factors beyond the farmers' control.

Conversely, the estimated value of sigma square (σ^2) is nearly negligible and holds statistical significance at the 1 per cent level. This shows that sigma has a good fit and assumptions of composite error term distribution were specified correctly. From the table, the estimated sigma parameter (σ^2) shows that about 2.6 per cent of the variation in fish production among the farmers was attributed to differences in the technical efficiencies of the fish farmers.

The variables, namely labour and organic fertilizers, displayed considerable significance and positively influenced fish production. Consequently, a 1 per cent increase in labour could potentially elevate fish production returns by 18 per cent, assuming all other factors remain constant. Similarly, a 1 per cent increase in organic fertilizer (i.e. cow dung) could increase the return from fish production by 50 per cent. From the table, the factor 'fish fingerlings' is highly significant at a 5 per cent level and has a positive impact on it. If the fingerling of fish is increased by 1 per cent, total production could be increased by 23 per cent. According to Devi et al. (2014), expenditures on fish fingerlings were identified as the crucial element in fish farming in Manipur. Therefore, it is imperative to prioritize investment in fish fingerlings. The factor, i.e., land area is significant at a 10 per cent level and has a positive impact on the production of fish. If it is increased by one unit, the total production could be increased by 32 per cent.

Factors Contributing to Technical Inefficiency in Fish Production

The determinants influencing technical inefficiency in fish production are outlined in Table 15.2. Given the inherent risks and uncertainties associated with fish farming, it is crucial for farmers to possess adequate skills and expertise to navigate the challenges effectively. Proficiency in various aspects such as scientific pond preparation, monitoring water quality parameters like pH and dissolved oxygen levels, understanding stock composition, selecting quality seedlings, managing stock density, providing quality feed and disease control are essential for successful fish farming endeavours. However, many farmers in the study area lack these technical skills, leading to reduced production and overall TE. Analysing technical inefficiency offers valuable insights into the factors contributing to inefficiencies in fish production. As noted by Giroh & Adebayo (2007), disparities in farmers' TE may stem from their individual characteristics and the prevailing technology. In this study, three variables – age, experience in fish farming and family size – were determined to be statistically insignificant in influencing technical inefficiency.

Table 15.2. Frequency Distribution of the Technical Efficiency of Fish Producers.

Technical Efficiency (per cent)	Frequency	Per cent to Total	Cumulative Per cent
0.80–0.85	1	1.54	1.54
0.85–0.90	7	10.77	12.31
0.90–0.95	27	41.54	53.85
0.95–1.00	28	43.08	96.92
Equal to 1	2	3.08	100.00
Minimum	0.83		
Maximum	1.00		
Mean	0.94		

Source: Self-computation from field survey, 2022.

Arrangement of Farms (or Fishermen) Based on Technical Efficiency Levels

The stochastic production frontier model offers a valuable advantage by enabling the estimation of technical efficiencies for each individual farm. Examination of Table 15.2 reveals that TE spans from 83 to 100 per cent for the least and most efficient farmers, respectively, with an average efficiency of 94 per cent. This suggests that, on average, fish farmers in the study area were operating below the maximum potential production level, experiencing a deficit of approximately 6 per cent. Notably, similar mean technical efficiencies above 70 per cent in carp production were also observed in studies conducted by Itam et al. (2015) and Baruwa & Omodara (2019).

Out of 65 samples, 55 samples (i.e., about 84 per cent) were in the range of 90–100 per cent. So, most of the farmers are very close to the production frontier. Only 3 per cent were considered the best farmers, whose TE score is equal to one.

Concluding Remarks and Policy Implications

It is true that North Bengal has abundant natural endowments and human skills in respect of intensified fish farming. But in spite of enormous potential, much ahead has not been made. The levels of production and productivity have not been adequate, resulting in a large gap between actual and potential output. The new technology has failed to produce the potential output. This may be due to a shortage of quality seeds, inadequate training programmes, unutilized water resources, technology, etc. As a consequence, it can be stated that present unsatisfactory performance will lead to the absence of rational management and hinder the growth of inland fishery in this region. Also, there is a need to change the attitude and motivation level of fish farmers since most of the fish farmers are in the business of living rather than maximizing the production of fish regardless of cost. Most of the fish farmers are risk-averse to being economically poor. Proper training with proper tools should be provided to the fishermen. Some sort of effective insurance for fish crops opens the door for the development of fish farmers. Therefore, attention should be given to boosting aquaculture production in North Bengal where fisheries is an important sector of employment as well as nutrition for low–income rural populations. This indicates that fish farmers have significant opportunities to boost their income and, consequently, alleviate poverty. Enhancements can be realized by minimizing inefficiency determinants to their optimal levels. Therefore, there is considerable room for increasing productivity substantially. Larinsangpui et al. (2016) noted that estimated efficiency and inefficiency indexes of sample farms may vary across different studies due to factors such as collected databases, location, time frames, farm structures and other variables.

Based on the findings of this study, it can be inferred that fish farmers in the surveyed region favour composite fish production, with rohu, mirgal and grass carp being the primary components. Augmenting investments in labour, organic fertilizers, fish fingerlings and land area could potentially enhance fish production output. Given the limited institutional sources of fish fingerlings in the area, establishing high-quality hatcheries and fingerling production centres becomes imperative. Furthermore, promoting the use of improved fish feed is essential for boosting production and maximizing returns. Implementing an organized farming system, such as a fish cooperative or producer's company, is crucial for facilitating informed decision-making and capitalizing on economies of scale in production. Additionally, fostering robust extension linkages and facilitating access to institutional credit are vital mechanisms for empowering farmers to adopt scientific farming techniques and improve TE.

These are the suggestions for the fish farmers that they should be economically as well as technically optimal since most of the fish farmers are in the business of making a living rather than trying to maximize fish production, regardless of the cost. Again, it is worth noting that being poor, small fish farmers tend to be risk-averse. They really trade off lower risk against higher profit or the disutility of increasing risk. So, there should be some sort of insurance for the fish crops, which pave the way for more adoption of these practices.

Fallow, derelict water bodies and waterlogged areas should be developed for fish culture. Credit institutions should be galvanized to simplify appropriate credit. Credit institutions are required to simplify the procedure for the disposal of loan applications. For reducing defaults in the repayment of loans, an effective post-disbursement monitoring system should be undertaken. Efforts should be made to provide institutional support to train fish farmers in order to enrich and motivate them in the field of scientific fish farming. The trainers should be periodically exposed to new knowledge and skills, and then the training they impart is then expected to be meaningful in the training. Aquaculture needs to be integrated with agriculture, piggery, duckery, etc. Nowadays, pisciculture activity is mixed with piggery or duckery activity for better viability of the pisciculture scheme. The product of the piggery or poultry may be utilized for manuring the water of the ponds or tanks. These will provide basic nutrients to the water, which also helps in maintaining the food chain of the fish. No extra expenditure for manure or supplementary feed will be required.

References

Aigner, D., Lovell, C., & Schmidt P. (1977). Formulation and estimation of Stochastic Frontier Production function models. *Journal of Econometrics*, *6*(1), 21–37, https://doi.org/10.1016/0304-4075(77), 90052–5.

Alam, M. A., Rahman, K. M. M., & Quddu, S. (2005). Measurement of economic efficiency of production fish in Bangladesh with translog stochastic cost frontier. *Bangladesh Agricultural Economists*, *XVIII*, 1–2.

Asamoah, D., Annan, J., & Nyarko, S. (2012). AHP approach for supplier evaluation and selection in a pharmaceutical manufacturing firm in Ghana. *International Journal of Business and Management*, *7*(10). DOI:10.5539/ijbm.v7n10p49

Baruwa, O. I., & Omodara, O. D. (2019). Technical efficiency of aquaculture system in Oyo State, Nigeria: Stochastic frontier approach. *Journal of Aquatic Research & Marine Sciences*, *2*(1), 114–120.

Coelli, T. J., Rao, P., & Battese, G. E. (1998). *An introduction to efficiency and productivity analysis* (pp. 263–288). Kluwer Academic Press, Boston, MA. https://doi.org/10.1007/978-1-4615-5493-6

Devi , S. Z., Singh, N. R., Singh, N. A., & Laxmi, T. H. (2014). Fish production in Manipur- an economic analysis. *Journal of Crop and Weed, 10*(2), 19–23.

Ekramuddin, M. (2017). An overview of impact of small scale fish farming on socio-economic growth in Birbhum district, West Bengal, India. *Journal of Entomology and Zoology Studies, 5*(4), 46–52.

Fare, R., Grosskopf, S., & Lovell, K. (1985). *The measurement of efficiency of froduction.* Kluwer-Nijhoff Publishing, Boston.

Giroh, D. Y., & Adebayo, E. F. (2007). Analysis of the technical efficiency of rubber tapping in Nigeria. *Journal of Agriculture and Social Sciences, 3*(4), 140–142.

Grafton, R., Arnason, R., Bjornal, T., & Campbell, D. (2006). Incentive -based approach to sustainable fisheries. *Canadian Journal of Fisheries and Aquatic Sciences, 63,* 699–710.

Greene, W. H. (1990). A Gamma-distributed stochastic frontier model. *Journal of Econometrics, 46*(1-2), 141–163.

Holborn, R., Baum, J., & Branch, T. (2009). Rebuilding global fisheries. *Science, 325*(5940), 578–585.

Hossain, M., Alam, A., & Uddin, K. (2015). Application of stochastic frontier production function on small banana growers of Kushtia District in Bangladesh. *Journal of Statistics Applications & Probability, 4*(2), 337–342.

Itam, K. O., Ajah, E. A., Ofem, U. I., & Ewa, O. (2015). AbamTechnical efficiency analysis of small scale Cassava Farmers in cross river state, Nigeria: A stochastic production frontier approach. *Applied Economics and Finance, 2*(4), 10–18.

Jondrow, J., Lovell, C., Materov, I., & Schmidt, P. (1982). On the estimation of technical inefficiency in the stochastic Frontier production function model. *Journal of Econometrics, 19*(2-3), 233–238.

Kirkley, J., Squires, D., & Walden, J. (1999). *Assessing efficiency and capacity in fisheries.* Prepared for the National Marine Fisheries Service workshop.

Kumbhakar, S., & Wang, H. (2010). Estimation of technical inefficiency in production Frontier models using cross-sectional data. *Indian Economic Review, 45*,(2), 7–77.

Kundu, S. K. (2015). *Inland fishing activities in Kumarganj Block of Dakshin Dinajpur*, West Bengal, India.

Majumdar, T. (2017). *Economics of small plantation in West Bengal – A study of production, efficiency and productivity performance using DEA and stochastic frontier regression*, PhD Dissertation, Department of Economics, University of North Bengal, India.

Meeusen, W., & van den Broeck, J. (1977). Efficiency estimation from cobb-Douglas production functions with composed Error. *International Economic Review, 18*(2), 435–444.

Misra, J., & Misra, S. R. (2014). Technical efficiency of fish farms in West Bengal: Nature, extent and implications. Agricultural Economics Research Review, Agricultural Economics Research Association (India), *27*(2), 221.

Moges, A., Beyene, A., Triest, L. Ambelu, A., & Kelbessa, E. (2018). Imbalance of ecosystem services of wetlands and the perception of the local community towards their restoration and management in Jimma Highlands, Southwestern Ethiopia. Wetlands, *38*, 1081–1095. https://doi.org/10.1007/s13157-016-0743-x

Olaniyi, O. (2012). Efficiency of fish production: An application of stochastic production function in Oyo Agricultural Zone, Oyo State, *Nigeria. International Journal of Applied Research and Technology, 1*(3), 53–62.

Pascoe, S., Kirkley, J. E., Greboval, D., & Morrison-Paul, C. J. (2003). Measuring and assessing capacity (p. 130). FAO Fisheries Technical Paper. No. 433/2.

Roy, T. (1999). Traditional industry in the economy of colonial India. Cambridge University Press, Cambridge.

Saha, S. (2016). *Economics of fishery and livelihood: A study in the coastal regions of West Bengal*, PhD Dissertation, University of Burdwan, India.

Singh, K., & Devi, C. (2020). Technical efficiency of small scale carp production in Manipur: A stochastic frontier approach. *Journal of Crops and Weed*, *16*(3), 160–165.

Umamaheswari, T., Rajakumar, M., & Chidambaram, P. (2017). Economics of ornamental fish farming industry in Madurai District, Tamil Nadu. *Indian Journal of Fisheries*, *64* (Special Issue), 273–277.

Worm, B., Hilborn, R., Baum, J. K., Branch, T. A., Collie, J. S., Costello, C., Fogarty, M. J., Fulton, E. A., Hutchings, J. A., Jennings, S., Jensen, O. P., Lotze, H. K., Mace, P. M., McClanahan, T. R., Minto, C., Palumbi, S. R., Parma, A. M., Ricard, D., Rosenberg, A. A., … Zeller, D. (2009). Rebuilding global fisheries. *Science*, *325*, 578–585. DOI:10.1126/science.1173146.

Chapter 16

Food Security and Safety in Bangladesh: Implications of Environmental Health

Md. Juel Mia[a], *Md. Shihabul Hossain*[a], *Shamima Prodhan*[a] *and Md. Saifullah Akon*[b]

[a]*Institute of Disaster Management and Vulnerability Studies, University of Dhaka, Bangladesh*
[b]*Department of Japanese Studies, University of Dhaka, Bangladesh*

Abstract

The geographical location, soil texture, and other environmental ingredients have made Bangladesh a bliss for several types of food production. Certain collapsible endeavours have made the potential agriculture sector less lucrative in the country. Even though food is the ultimate purchasing goal to remain alive and healthy, there have been empirical shreds of evidence that indicate Bangladesh has been going through vulnerable phases ensuring solid food security and safety. The chapter explores Bangladesh's vulnerability to environmental stressors and climate change impacts that significantly affect food availability and agricultural productivity highlighting the pervasive issue of food safety in Bangladesh and identifying contaminants such as pesticide residues as a major threat to public health and food security. Then it focuses on the country's huge potential to ensure technology-driven agriculture production that may further strengthen food security and the economy with a special focus on the governance and environmental issues in ensuring sustainable agriculture. From the worldview, the government of Bangladesh needs to adopt a policy that will ensure the national interest and the safety of the citizens inside the country which will help to avoid any devastating food crisis during emergencies and international disputes.

Keywords: Sustainable agriculture; food security; food safety; public health; environmental hazards; Bangladesh

Sustainable Agricultural Practices: Economic and Environmental Implications, 237–250

doi:10.1108/978-1-83608-336-820241016

Introduction

Food is an essential element for sustaining life; it helps curb hunger and provides nutrient support and energy. We noticed food insecurity in different parts of the world as well as the huge amount of food loss. There are four basic dimensions of food security such as availability, access and choice, utilization, and stability (FAO, 2006). Reviewing relevant literatures, Manikas et al. (2023) illustrated that in the research, household-level calorie adequacy indicator has been used as sole measure of food security; dietary diversity-based and experience-based indicators were used frequently; however, utilization and stability dimensions were rarely found. Timmer (2000) explained three marco dimensions of food security such as economic growth, equitable distribution and food price stability. Development in agriculture and rural growth largely influenced by government policies and also characterized by the outcome of rest of the economy. Author suggested that strategies including 'growth with redistribution ... encourages balanced growth to create non-farm employment opportunities, while investing in rural infrastructure'. World Food Summit held in 1996 defined food security as 'all people, at all times, have physical and economic access to sufficient, safe and nutritious food that meets their dietary needs and food preferences for an active and healthy life' (FAO, 2006).

Obtaining safe food is basic human rights and it helps safeguarding individual and public health and ensures sustainable economic growth. However, millions of people in the world are at risk due to unsafe food (Fung et al., 2018). To provide healthy life, One Health approach should get attention and implementation. This concept interlinks health of humans, animals and environment and 'mobilizes multiple sectors, disciplines, and communities at varying levels of society to work together to foster well-being and tackle threats to health and ecosystems' (Adisasmito et al., 2022) and 'can be applied to food safety, sustainable food production, and environmental stewardship by bringing together interdisciplinary teams to create a One Health network to address these challenges' (Garcia et al., 2020). Variagated aspects are related to safe and healthy food such as chemical, biological, personal hygiene and environmental issues (Fung et al., 2018). Due to the industrial pollutants, historically, food is contaminated in every countries which is true from developed to poor countries. For producing more foods farmers are using pesticides and chemical fertilizers, for preserving food different chemical ingredients are used, for packing purposes polythene and other food containers are used. From producers to consumers and from business enterprises to government agencies have responsibilities to ensure food safety thus to contribute human health. This chapter focuses on the food security and safety in Bangladesh linking environmental health aspects. Furthermore, this writing tries to explain the ethical responsibilities toward obtaining food security and safety for present generation and for the next generations.

In the context of Bangladesh's geographical setting, the land exhibits considerable productivity and potential within the agricultural sector. Influential factors such as soil type, quality, river paths, weather conditions and the expertise of the human population contribute collectively to the heightened potential for agricultural production in the region. In recent years Bangladesh has been doing outstanding performance particularly increases the agricultural production though the agricultural lands are decreasing in an alarming rate. On the other hand, Bangladesh faces a multifaceted landscape of challenges, where environmental issues, development politics and climatic hazards perspectives converge. The impacts of climate change and environmental degradation loom large, necessitating strategic responses to this dynamic and evolving scenario. The nation contends not only with the direct consequences of these environmental shifts but also grapples with the ripple effects of unplanned land use and a lack of innovation. Limited progress in adopting scientific advancements further complicates the complexity of ensuring a reliable and secure food supply.

Within this challenging landscape, the allocation of land for diverse food production in rural areas emerges as a critical focal point. Strategic planning becomes imperative to mitigate dependence and avert shortages, demanding a comprehensive understanding of the intricate dynamics involved in achieving food safety and security in the country. Bangladesh's vulnerability to climate-related risks adds an additional layer of complexity, as evident from reports by (Mainuddin & Kirby, 2015) citing threats such as floods, droughts, salinity and rising sea levels. Recognizing the gravity of these intertwined challenges, the National Food Policy, articulated by the Ministry of Food and Disaster Management of the Government of Bangladesh in 2006, underscores food security as a paramount concern. The policy outlines a central objective to ensure consistent food security for all citizens under diverse circumstances. As this chapter delves into the implications of environmental health on food security and safety in Bangladesh, the aim is to contribute to a deeper and more comprehensive understanding of the intricate dynamics at play in this critical aspect of the nation's development.

Food Security in Bangladesh

Bangladesh has experienced sustained economic growth, making it one of Asia's fastest-growing economies. However, despite this progress, 11% of Bangladeshi households still experience food insecurity (WFP, 2023). Bangladesh, with a population of about 174 million people (World Population Review, 2024) and a land area of 148 million hectares (Mha), has a particularly pressing challenge with food security. After nine-month liberation war Bangladesh gained independence from Pakistan on 16 December 1971. In the initial stage, almost 80% of people of the country lived below the poverty line, and the poverty rate significantly decreased over time and now the percentage is 20. However, the number of populations living under the poverty line not yet decreased; according to the Bangladesh Poverty Watch Report 2022 about 35 million people live below the poverty line (Mavis, 2023). Overpopulation, along with the decrease in the land-to-human ratio, has made food security of utmost necessity. Moreover, Bangladesh's low GDP (US$2,687/year) (MCCI, 2023) and high population density (1,342/km^2; Worldometer, 2024) make food security paramount for the country's development.

Bangladesh is listed in the Climate Risk Index 2021 as one of the country's most adversely affected by climate change. Its susceptibility to food insecurity has increased due to shifting patterns of natural disasters and hazards including floods, irregular and unexpected rainfall, temperature fluctuations, rising salinity, untimely droughts, river erosion and earthquakes (Sultana & Sabau, 2023) by impacting lives and livelihoods of people and economy of the country. The Ministry of Agriculture estimates that the recent widespread floods damaged crops valued at about BDT 13 billion (USD 151.672 million), submerging 159,000 acres of agricultural land (Hossain, 2020). According to the Ministry of Agriculture estimation, storm Amphan (2020) damaged crops on 176,007 acres, resulting in a total crop loss of USD 6.72 billion over 17 coastal districts (Dhali, 2020; Wardad, 2020). Bangladesh has a long-standing tradition of preparing for and responding to agricultural disasters. Farmers adapt their fragmented landholdings and cropping patterns to the specific potentials and risks of various land and soil types. The recent expansion of irrigation has significantly boosted crop production during the dry season. Sea level rise is another threat that could adversely affect coastal areas of the country (Brammer, 2020). According to World Bank projections, sea levels would rise by 0.1, 0.25 and 1 m by 2020, 2050, and 2100, respectively, affecting 2, 4 and 17.5% of the land area (GFDRR, 2011). As an agricultural country with limited available land (less than 0.1 ha per capita), agriculture contributed around 11.50 per cent of the total GDP. It employed 45.4% of the workforce in FY 2021–2022 (BBS, 2023). Likewise, the livestock and fisheries subsector significantly influence Bangladesh's economic growth and food security. In Bangladesh, the livestock industry directly employs 20% of the labour force and indirectly employs 50% of the labour force. Concurrently, the aquaculture industry is nearly twice as large as the livestock industry, accounting for 26.37% of the overall agricultural GDP and 3.52% of the national GDP (Ministry of Fisheries and Livestock, 2021).

The concerns of food security, livelihoods and economic growth are all addressed by agriculture, making it one of the most significant economic sectors in Bangladesh. However, the country's food security is in jeopardy due to scarce arable land, inadequate irrigation water and severe weather. Rapid population growth in Bangladesh has decreased the per-capita agricultural land: 0.11, 0.09, 0.07 and 0.06 ha/person in 1981, 1991, 2001 and 2011, respectively (Rai, Zhang, Paudel, Li, & Khanal, 2017). Bangladesh is currently experiencing pressure of urban expansion and population growth that significantly contributes to the reduction of available arable land. Between 1976 and 2010, total agricultural land – including cropland, forest, mangrove, river, lake, beel (lake-like stagnant water body), hair (bowl-shaped shallow depression), aquaculture, tea estates and saltpans – decreased by 1.12 million ha (Hasan et al., 2013) and 1.22 million ha more non-agricultural land had been added (Rai et al., 2017). On the other hand, unusual weather patterns, particularly flash floods, and its erratic climate significantly impact the nation's crops, especially rice production. Degraded soil, poor resource distribution and reliance on monsoons exacerbate the problem. Furthermore, sea level rise 7.8 mm annually along Bangladesh's coastline, compared to an average of 3 mm globally. Nevertheless, Bangladesh has also achieved significant progress in ensuring food security over the last 40 years despite its fast population increase and susceptibility to natural calamities. For example, between 1972 and 2014, the production of food grains tripled from approximately 10 million to 34 million tons (The World Bank, 2016). The country also made notable progress in rice production, which is the country's staple food. In 1971, rice production was 10.6 million tons, which tripled to 36.3 million tons in 2020 (Hassan, 2021). Similarly, wheat and maize production have also steadily increased over the last few decades. The annual production of potatoes has increased substantially from 1 million tons in the mid-1980s to the present level of 10 million tons. There have also been increases in the production of oil-seed (450,000 tons in 1987/1988 to more than 1,000,000 tons in 2019), freshwater fisheries (growth of 5 percent per annum over the last two decades) and livestock (1.5 percent contribution to the total GDP) (BBS, 2019).

In Bangladesh, agricultural land is being converted to non-agricultural land (e.g. houses, roads and business enterprises), which may have worsened the effects on food security (Quasem, 2011). Other factors contributing to the loss of arable land include deforestation, excessive rainfall, strong river currents and silt deposition. According to IPCC projections, Bangladesh's wheat and rice production is expected to fall 8–32% due to climate change by 2050 (OECD, 2009). Furthermore, the same analysis predicts that by 2050, floods and saline intrusion will cause Bangladesh's pre-monsoon and dry-season rice yield to drop. According to the WFP (2023) report, 68% of households were significantly affected by food prices, with a significant increase in most food commodities compared to the last year (2022). Prices of essential foods have been increasing almost every day in Bangladesh since the late 2007 food crisis. Additionally, the fragility of the world's food supplies is shown by the gradual increase in food prices in 2004 and the significant fluctuations in 2024. Extreme weather conditions can also deteriorate the infrastructure for food delivery and storage, causing food to decay and go to waste. All these elements contribute to Bangladesh's rising food security dilemma, which is expected to worsen as climate change worsens.

The government of Bangladesh has made great efforts to curb population growth and combat food insecurity since the country's independence in 1971 (Parvin & Ahsan, 2013). Bangladesh has had a national food security strategy since 2006, with most of its household provisions. As part of its food security initiatives, the Bangladeshi government has started open market sales (OMS), free food distribution, a food-friendly programme, various alleviation operations, etc. These programmes aim to guarantee that every citizen has sufficient access to wholesome meals. To address economic vulnerability of the marginalized people, the government of Bangladesh is implementing several safety net programmes. To protect vulnerable people from food insecurity, the government introduced a social safety net programme in 1998. In FY 2019, a budget of approximately BDT 642 billion, or equivalent to 2.5 percent of the gross domestic product (GDP), has been allocated for this purpose, which makes up almost 14 percent of the national budget and covers 27 percent of households in the country. The Text Relief programme is one of Bangladesh's oldest social safety net programmes. It primarily began as a relief programme in 1975 in response to the famine of 1974. Its immediate objective was to provide food to the severely food-insecure poor. The Vulnerable Group Feeding (VGF) programme was launched after the 1974 famine in partnership with the World Food Programme (WFP). Initially, VGF provided a monthly transfer of 31.25 kg of wheat per household for two consecutive years. As of FY19, the budget allocation represents 2.7 percent of the social protection budget and 0.07 percent of GDP. The average benefit per card (beneficiary) per annum has hovered around BDT 13,302. The Food for Work (FFW) programme in Bangladesh has been a crucial safety net, providing employment opportunities and food assistance to those in need. Instead of cash wages, workers were paid in food grains under this programme. The FFW programme was launched in 1974 during a severe food shortage in the country. With its nationwide operation, its direct beneficiaries include approximately 750,000 individuals. Over US$140 million, making it one of Bangladesh's most extensive safety net programmes. VGF programme provides nutritious meals to improve the health and well-being of marginalized groups, including pregnant women, lactating mothers and children. The VGF programme primarily began as a relief effort in 1975, which aims to provide food assistance to the severely food-insecure poor during lean periods. When introduced, VGF provided a monthly transfer of 31.25 kg of wheat per household for two consecutive years. OMS aims to provide food security to relatively poor households during various situations, including natural disasters and economic challenges. The OMS programme has been in effect since the early 1980s in Bangladesh. OMS sells subsidized essential food items in local markets to stabilize prices and ensure affordability for low-income households. Under the OMS programme, approximately one crore beneficiaries receive rice at Tk 30 per kg, while 50 lakh particular cardholders obtain rice at Tk 15 per kg (The Daily Star, 2022). However, the recent seasonal OMS, a subsidized rice scheme, was first introduced in April 2008. It was initiated in response to rising food costs due to global price hikes.

Perspectives on Sustainable Agriculture

The idea of sustainable agriculture is ambiguous and can be seen from different perspectives, e.g., economics and business, environmental studies, social sciences and governance. A report 'Our common future' by the world commission of environment and development was published in 1987 which explained the necessities of applying the concept of sustainability in different fields and it successfully attracted attention of world communities. After the publication of the report, the concept of sustainability has been applied in the variegated fields of agriculture. The report described that 'growth in world cereal production has steadily outstripped world population growth. Yet each year there are more people in the world who do not get enough food. Global agriculture has the potential to grow enough food for all, but food is often not available where it is needed' (United Nations, 1987). The U.S. Farm Bill in 1990, described as an 'integrated system of plant and animal production practices', outlines the principles of sustainable agriculture. It aims to achieve various long-term objectives including meeting human food and fibre needs, enhancing environmental quality, utilizing resources efficiently, maintaining economic viability and improving the overall quality of life for farmers and society. Importantly, this approach must consider environmental, social and economic sustainability, reflecting the three fundamental pillars of sustainable development.

The Food and Agriculture Organization (FAO) provided a clear definition of sustainable agriculture in 1988, describing it as a careful synthesis of conservation and natural resource management. The crux of the matter is the deliberate synchronization of institutional and technical advancements with the goal of ensuring that human needs are met for present and future generations. The essence lies in the strategic alignment of technological and institutional changes, aiming to guarantee the fulfilment of human needs for both current and future generations. This definition intricately weaves together the environmental, technological and institutional facets, highlighting their interconnectedness.

Within this paradigm, the primary objective is the preservation of crucial elements such as land, water and the genetic resources of plants and animals. The essence of sustainability, as per the FAO, lies in strategically aligning technological and institutional changes to ensure the fulfilment of human needs for both present and future generations, reflecting a forward-looking perspective. The interconnected environmental, technological and institutional facets are described as intricately woven together, highlighting the necessity of a holistic approach for sustainable agricultural practices (Wijerathna-Yapa & Pathirana, 2022).

The discourse surrounding sustainable agriculture has raised concerns about a potential imbalance, with a predominant focus on the lower levels of the continuum while overlooking the higher levels (Cook et al., 2015). For attaining sustainability in agriculture production and distribution it requires comprehensive and holistic approach in sustainable agriculture, taking into account not only immediate environmental and economic considerations but also the broader implications for societal well-being and the attainment of sustainable development goals.

Sustainable agriculture is essential as it tackles environmental, economic and social aspects crucial for maintaining a balanced ecosystem. Its significance lies in its ability to counteract issues like soil degradation, deforestation and water contamination, which are outcomes of traditional farming methods. Sustainable agriculture prioritizes the preservation or enhancement of soil fertility, reduction of water consumption and protection of biodiversity to ensure the prolonged health of ecosystems (Kremsa, 2021). With the global population on the rise, the demand for food production has surged, making efficient resource utilization crucial. Through practices like crop rotation and agroforestry, sustainable agriculture aims to minimize waste and conserve essential resources such as water and arable land, ensuring responsible resource utilization without compromising the needs of future generations. Additionally, it plays a crucial role in addressing climate change by advocating for techniques like agroecology and organic farming, contributing to carbon sequestration and reducing greenhouse gas emissions (Sinclair et al., 2019). The economic sustainability of farming operations is improved by embracing sustainable practices, promoting crop diversification, enhancing soil health and decreasing dependence on external inputs. This economic stability is vital for the livelihoods of farmers and bolsters the overall resilience of agricultural systems.

Sustainable agriculture also aids in the preservation of biodiversity by advocating for diverse cropping systems and habitats, countering the species loss associated with monoculture. It addresses concerns about water quality by emphasizing responsible water management and reducing the use of harmful chemicals. By contributing to long-term food security, sustainable agriculture enhances the adaptability of farming systems to changing conditions, reducing vulnerability to pests, diseases and climatic variations (Farooq et al., 2019). This emphasizes the vital importance of sustainable agriculture in securing food in the long term. It suggests that practices fostering stability and reliability in food production over time are essential for ensuring a steady food supply. Sustainable agriculture is portrayed as a means of improving the adaptability of farming systems, suggesting that it offers the necessary flexibility to respond effectively to changing conditions. Furthermore, the noted decrease in vulnerability to pests, diseases and climatic variations underscores the objective of sustainable agriculture to build resilient farming systems capable of enduring and mitigating potential threats. This connection between sustainable agriculture and adaptability suggests that the practices not only foster environmental sustainability but also contribute to the overall resilience and robustness of agricultural systems. Overall, the statement conveys a holistic perspective, emphasizing that the benefits of sustainable agriculture extend beyond immediate production concerns, addressing broader challenges related to food security and the adaptability of farming practices. Furthermore, it promotes social well-being by endorsing fair labour practices, encouraging community engagement and ensuring equitable resource distribution. This empowerment of farmers supports rural economies and enhances the overall welfare of communities.

Certainly, the in-depth conversation about sustainable agriculture in Bangladesh reveals a complex and evolving situation driven by the need to address urgent challenges. The country is dedicated to balancing the increasing demand for food with environmental and social considerations. The adoption of integrated farming practices, organic farming and water management techniques reflects a comprehensive approach to using resources wisely and conserving biodiversity (Muhie, 2022). Importantly, the involvement of farmers in community-based initiatives strengthens the impact of sustainability at the grassroots level, demonstrating a collaborative effort toward a robust agricultural sector (Farooq, Rehman, & Pisante, 2019). The focus on climate-resilient crops, government initiatives and educational programmes highlights proactive measures taken to tackle the challenges posed by climate change. Examining specific practices like rice-fish farming systems, biogas production and women's involvement provides additional insights into how sustainability is woven into various aspects of Bangladeshi agriculture. Incorporating technology, research and international collaborations signals a forward-thinking approach, while the attention to waste reduction and recycling underscores a commitment to a circular and sustainable agricultural system. In essence, Bangladesh's comprehensive strategy represents a united effort to develop a sustainable and resilient agricultural sector that not only ensures food security but also preserves the country's natural resources for future generations.

Environmental Health and Moral Responsibilities

Environmental health does not just explain the health of the environment, but it also refers to the environmental aspects that affect human health. Environmental health largely depends on the relationship between humans and the environment/nature, and the concept promotes policies to reduce impacts of natural and human-induced environment on the individual and population health. Though there are different points of view and debate considering the relative significance of human interests, however, environmental ethics is concerned with the moral relationship between human and nature, intrinsic value of nature and human responsibilities toward nature and the environment (Aceves-Avila, 2020; Shrader-Frechette, 2009).

Bangladesh is a densely populated country and most population largely depend on agricultural activities. To increase crop production farmers depend on chemical fertilizers and pesticides which are detrimental to public health. Another issue is that in the environmental settings and agricultural products, unregistered pesticides are found that represent weakness in the pesticide governance. In the intricate tapestry of sustainable agriculture, the interconnection between food ethics and public health takes centre stage, shaped significantly by conscientious agricultural practices with a core focus on sustainability. These ethical considerations serve as a linchpin, exerting a direct impact on public health outcomes through a delicate interplay among environmental, economic and social dimensions. Sustainable agriculture functions as a moral guidepost directing efforts towards enhanced nutritional security, reduced chemical exposure and the preservation of water quality, thus addressing malnutrition and promoting overall well-being. Ethical commitments extend to climate change mitigation, social equity and fair trade, positively influencing the well-being of farmers and communities. Local and traditional food preservation, transparency, support for biodiversity and inclusive decision-making are integral to ethical sustainable agriculture, fostering community well-being and resilience. The prevention of exploitative labour practices, responsible land use and mitigation of antibiotic resistance align with public health goals. Considerations for community health, sanitation, consumer empowerment, crisis preparedness and holistic health promotion contribute to a comprehensive ethical framework.

Researchers suggest that primarily the producers, distributors, handlers and vendors are responsible for healthy and safe food; and at the same time the consumers should be aware of food safety (Fung et al., 2018). To protect individual and population health, the government should take initiative and enforce food safety regulations, and relevant stakeholders should be vigilant from production areas to shops. Reducing over consumption and wastage can significantly address food security and safety. Proper distribution and management, good governance, setting up food banks and voluntary donations can help vulnerable people to access food and nutrition. As culture has relation with food, there is scope of considering culture into account in improving food security and ensuring safety (Alonso et al., 2018). Haider (2021) uttered that 'Time and again we have been reminded of the dire condition of food safety in Bangladesh. Although positive steps are being taken and some praiseworthy initiatives are in place, we still have a long way to go in ensuring safe food for all. The important thing to understand is that it is a collective responsibility'. Communicating environmental health risks and ethical responsibilities to the different segments of environmental management audiences can reduce risks; however, communicating environment, health and risk is challenging, no single approach works effectively, and different strategies can be taken based on the characteristics of the audiences (Fitzpatrick-Lewis et al., 2010).

The National Parliament of Bangladesh has passed and implemented food safety act in 2013, and following the act, Bangladesh Food Safety Authority was established in 2015. Still the act is not fully executed. In 2021, on the occasion of national food safety day, a rally was held in Dhaka where experts and volunteers were attended, and they urged full implementation of the act for ensuring proper action against food adulteration and to safeguard the present and future generations (TBS News, 2021). Though the magistrates occasionally visit food shops, food yards and manufacturing plants to inspect, however, the contemporary system for prevention of food adulteration failed to curb the common adulteration and malpractices in Bangladesh due to inactive surveillance programme for prevention of food adulteration (Rahman, 2017).

Food Security and Safety: Aid and Cooperation

Disasters, whether natural or man-made, have often led to food crises including food production, distribution and access that threaten the well-being of the affected population. These disruptions can lead to immediate food shortages and long-term food insecurity. In the aftermath of disasters, communities may find their agricultural systems crippled and infrastructure damaged. These factors contribute to a heightened risk of malnutrition and hunger. To mitigate the adverse effects of such food insecurity, international aid plays a crucial role. Humanitarian organizations, governments and non-governmental organizations (NGOs) collaborate to provide emergency assistance, including food supplies, medical aid and shelter. Immediate relief efforts became essential to address the urgent needs of affected populations, ensuring access to food, clean water and healthcare. International aid not only saves lives in the short term but also lays the groundwork for long-term resilience. International organizations, particularly the United Nations (UN), through agencies like the WFP and the FAO, play the central role in coordinating and delivering aid during food crises. These organizations work closely with governments, NGOs and the private sector to mobilize resources and expertise (FAO, 2017).

By situating in a cyclone-prone region, Bangladesh faces recurrent floods, cyclones and other calamities. These natural disasters often led to severe food crises. The fertile plains that contribute to the nation's agricultural productivity become vulnerable during floods and causing substantial damage to infrastructure. These calamities have a direct impact on food production and exacerbate food insecurity among vulnerable populations. One notable example is the devastating Cyclone Sidr in 2007, which left a trail of destruction in its wake. It affected over nine million people and destroyed crops, leading to a significant food crisis. The immediate response from the international community played a crucial role in providing emergency relief. After Sidr, the WFP collaborated with the Bangladesh government to provide emergency food assistance, reaching millions of people in the affected areas. This aid not only prevented immediate hunger but also laid the foundation for recovery efforts in addressing food crises.

The World Bank has actively supported Bangladesh's efforts to enhance food security and improve the livelihoods of vulnerable populations. Recently, the World Bank and the government of Bangladesh signed a $120 million financing agreement. This funding includes improving irrigated agricultural and fisheries production to increase the incomes of 170,000 poor people who are particularly vulnerable to the impacts of climate change (The World Bank, 2022). Urban Gardening tackles urbanization-related poverty and food insecurity. Low-income families can regularly access fresh, wholesome fruits and vegetables by cultivating their produce. The FAO has been actively involved in promoting urban gardening in Dhaka through partnerships with groups including BRAC, the Global Alliance for Improved Nutrition, Proshika, and the POCAA. Through its Dhaka Food System project, FAO has taught 400 women from underprivileged communities and given them tools, fruit saplings, vegetable seeds, fertilizer and other supplies. The Policy Research and Strategy Support Program for Food Security and Agricultural Development in Bangladesh was introduced by the International Food Policy Research Institute and regional partners. This programme, which USAID funds, attempts to address issues with hunger and malnutrition in the nation. Furthermore, USAID oversees Bangladesh's Feed the Future programmes, which emphasize raising production, enhancing food security and updating agricultural laws (CRI, 2019).

Conclusion

In developing countries, agriculture confronts a multifaceted landscape, balancing significant potential and productivity with challenges such as climate change, environmental degradation and the inadequacy to meet basic needs among rural agricultural communities, all posing threats to food security. The concept of sustainable agriculture, as articulated by the FAO, emerges as a promising solution that comprehensively addresses environmental, economic and social aspects for fostering a balanced ecosystem. This discourse emphasizes the urgency of adopting practices prioritizing soil fertility, reduced water consumption, biodiversity protection and economic sustainability. The critical role of educating farmers in techniques like crop rotation, agroforestry and organic farming is underscored, pivotal in minimizing waste, conserving resources and contributing to climate change mitigation, thereby bolstering adaptability to pests, diseases and climatic variations. Initiatives for sustainable agriculture represent a collective effort, integrating practices such as integrated farming, organic farming and water management. Active farmer participation in community initiatives, coupled with proactive measures like climate-resilient crops, government initiatives and educational programmes, signifies a commitment to addressing challenges posed by climate change to safeguard environmental health. The utilization of technology, research and international collaborations reflects a forward-thinking approach, while waste reduction and recycling illustrate dedication to a circular and sustainable system. Crucially, the nexus between sustainable agriculture, food ethics and public health is highlighted, influencing nutritional security, reducing chemical exposure and preserving water quality.

References

Aceves-Avila, C. D. (2020). Environmental ethics and justice for sustainable cities. In W. Leal Filho, A. Marisa Azul, L. Brandli, P. Gökçin Özuyar, & T. Wall (Eds.), *Sustainable cities and communities* (pp. 167–174). Springer International Publishing. https://doi.org/10.1007/978-3-319-95717-3_64

Adisasmito, W. B., Almuhairi, S., Behravesh, C. B., Bilivogui, P., Bukachi, S. A., Casas, N., Becerra, N. C., Charron, D. F., Chaudhary, A., Zanella, J. R. C., Cunningham, A. A., Dar, O., Debnath, N., Dungu, B., Farag, E., Gao, G. F., Hayman, D. T. S., Khaitsa, M., Koopmans, M. P. G., … Zhou, L. (2022). One health: A new definition for a sustainable and healthy future. *PLOS Pathogens, 18*(6), e1010537. https://doi.org/10.1371/journal.ppat.1010537

Alonso, E. B., Cockx, L., & Swinnen, J. (2018). Culture and food security. *Global Food Security, 17*, 113–127. https://doi.org/10.1016/j.gfs.2018.02.002

BBS. (2019). *Livestock economy at a glance [2018–2019]*. Department of Livestock Services. Retrieved March 6, 2024, from http://www.dls.gov.bd/site/page/22b1143b-9323-44f8-bfd8-647087828c9b/Livestock-Economy

BBS. (2023). *Labour force survey Bangladesh 2022*. Bangladesh Bureau of Statistics. Retrieved March 6, 2024, from https://bbs.portal.gov.bd/sites/default/files/files/bbs.portal.gov.bd/page/b343a8b4_956b_45ca_872f_4cf9b2f1a6e0/2023-10-25-07-38-4304abd7a3f3d8799fbcb59ff91007b1.pdf

Brammer, H. (2020). Agricultural hazard management in Bangladesh. *Agriculture for Development, 39,* 12–16. https://taa-international.org/wp-content/uploads/2020/04/Ag4Dev39_Spring_2020_web-1.pdf

Cook, S., Silici, L., Adolph, B., & Walker, S. (2015). *Sustainable intensification revisited*. International Institute for Environment and Development. Retrieved February 10, 2024, from https://www.iied.org/sites/default/files/pdfs/migrate/14651IIED.pdf

CRI. (2019). *Bangladesh: Towards achieving food security 2009-2019*. Center for Research and Information (CRI). Retrieved March 1, 2024, from https://cri.org.bd/publication/2019/Aug/food-security-19/Bangladesh%20-%20Towards%20Achieving%20Food%20Security%202009-2019.pdf

Dhali, B. (2020, May 21). *Cyclone Amphan: 1,76,007 hectare agricultural land affected*. Retrieved March 13, 2024, from *Dhaka Tribune*. https://www.dhakatribune.com/bangladesh/agriculture/210571/cyclone-amphan-1-76-007-hectare-agricultural-land

FAO. (2006). *Food security: Policy brief.* Food and Agriculture Organization. Retrieved March 14, 2024, from https://www.fao.org/forestry/13128-0e6f36f27e0091055bec28ebe830f46b3.pdf

FAO. (2017). *Global report on food crisis 2017*. United Nations Food and Agriculture Organization. Retrieved from https://www.wfp.org/publications/global-report-food-crisis-2017

Farooq, M., Rehman, A., & Pisante, M. (2019). Sustainable agriculture and food security. In M. Farooq, & M. Pisante (Eds.), *Innovations in sustainable agriculture* (pp. 3–24). Springer. https://doi.org/10.1007/978-3-030-23169-9_1

Fitzpatrick-Lewis, D., Yost, J., Ciliska, D., & Krishnaratne, S. (2010). Communication about environmental health risks: A systematic review. *Environmental Health, 9*, 67. https://doi.org/10.1186/1476-069X-9-67

Fung, F., Wang, H. S., & Menon, S. (2018). Food safety in the 21st century. *Biomedical Journal, 41*(2), 88–95. https://doi.org/10.1016/j.bj.2018.03.003

Garcia, S. N., Osburn, B. I., & Jay-Russell, M. T. (2020). One health for food safety, food security, and sustainable food production. *Frontiers in Sustainable Food Systems, 4*. https://doi.org/10.3389/fsufs.2020.00001

GFDRR. (2011). *Climate risk and adaptation country profile: Bangladesh*. Global Facility for Disaster Reduction and Recovery. Retrieved March 1, 2023, from https://www.gfdrr.org/en/publication/climate-risk-and-adaptation-country-profile-bangladesh

Haider, M. H. (2021, June 7). *Food safety is everyone's responsibility*. Retrieved February 5, 2024, from https://www.thedailystar.net/lifestyle/check-it-out/news/food-safety-everyones-responsibility-2106085

Hasan, M., Hossain, M., Bari, M., & Islam, M. (2013). Agricultural land availability in Bangladesh. *Dhaka: Soil Resource Development Institute (SRDI)*, Ministry of Agriculture.

Hassan, S. (2021, June 21). *The rice economy*. Retrieved March 15, 2024, from *The Daily Star*. https://www.thedailystar.net/business/economy/news/the-rice-economy-2114957

Hossain, I. (2020, August 19). *Flood costs Bangladesh crops worth 1,323C*. Retrieved March 13, 2024, from *Dhaka Tribune*. https://www.dhakatribune.com/bangladesh/220044/flood-costs-bangladesh-crops-worth-1-323c

Kremsa, V. Š. (2021). Sustainable management of agricultural resources (agricultural crops and animals). In C. M. Hussain, & J. F. Velasco-Muñoz (Eds.), *Sustainable resource management: Modern approaches* and contexts (pp. 99–145). Elsevier. https://doi.org/10.1016/B978-0-12-824342-8.00010-9

Mainuddin, M., & Kirby, M. (2015). National food security in Bangladesh to 2050. *Food Security*, *7*(3), 633–646. https://doi.org/10.1007/s12571-015-0465-6

Manikas, I., Ali, B. M., & Sundarakani, B. (2023). A systematic literature review of indicators measuring food security. *Agriculture & Food Security*, *12*(1), 10. https://doi.org/10.1186/s40066-023-00415-7

Mavis, M. (2023, January 22). *Report: 35m Bangladeshis still live below poverty line*. Retrieved March 13, 2024, from *Dhaka Tribune*. https://www.dhakatribune.com/business/281301/report-35m-bangladeshis-still-live-below-poverty

MCCI. (2023). *Bangladesh's economy during FY2022-23 (FY23)*. Metropolitan Chamber of Commerce & Industry (MCCI). Retrieved March 1, 2024, from https://mccibd.org/wp-content/uploads/2023/10/Bangladesh-Economy-FY-22-23.pdf

Ministry of Fisheries and Livestock. (2021). *Year book of fisheries statistics of Bangladesh 2019-20*. Department of Fisheries Bangladesh, Ministry of Fisheries and Livestock. Retrieved March 1, 2023, from https://fisheries.portal.gov.bd/sites/default/files/files/fisheries.portal.gov.bd/download/fa8a7744_bb11_4472_8afe_425083d221c8/2021-09-15-09-10-577655ff2458bff8c7377e82e7ed4c43.pdf

Muhie, S. H. (2022). Novel approaches and practices to sustainable agriculture. *Journal of Agriculture and Food Research*, *10*, 100446. https://doi.org/10.1016/j.jafr.2022.100446

OECD. (2009). *Integrating climate change adaptation into development co-operation*. OECD Publishing. Retrieved March 13, 2024, from https://www.oecd.org/environment/cc/44887764.pdf

Parvin, G. A., & Ahsan, R. S. (2013). Impacts of climate change on food security of rural poor women in Bangladesh. *Management of Environmental Quality*, *24*(6), 802–814. https://doi.org/10.1108/MEQ-04-2013-0033

Quasem, M. A. (2011). Conversion of agricultural land to non-agricultural uses in Bangladesh: Extent and determinants. *The Bangladesh Development Studies*, *34*(1), 59–85. Retrieved from https://www.jstor.org/stable/23342764

Rahman, B. (2017, August 29). *Food safety in Bangladesh: Challenges and concerns*. Retrieved February 1, 2024, from *Daily Sun*. https://www.daily-sun.com/post/251275/Food-Safety-in-Bangladesh:-Challenges-and-Concerns

Rai, R., Zhang, Y., Paudel, B., Li, S., & Khanal, N. (2017). A synthesis of studies on land use and land cover dynamics during 1930–2015 in Bangladesh. *Sustainability*, *9*(10), 1–20. https://doi.org/10.3390/su9101866

Shrader-Frechette, K. (2009). Environmental ethics. In H. LaFollette (Ed.), *The Oxford handbook of practical ethics* (pp. 188–216). Oxford University Press. https://doi.org/10.1093/oxfordhb/9780199284238.003.0009

Sinclair, F., Wezel, A., Mbow, C., Chomba, S., Robiglio, V., & Harrison, R. (2019). *The contribution of agroecological approaches to realizing climate-resilient agriculture*. Global Center on Adaptation, Rotterdam. Retrieved February 16, 2024, from https://gca.org/reports/the-contributions-of-agroecological-approaches-to-realizing-climate-resilient-agriculture/

Sultana, N., & Sabau, G. (2023). Food policy and food security in Northern Bangladesh: An empirical study of the impact of food policy on household food security. *Social Sciences & Humanities Open*, *8*(1), 100508. https://doi.org/10.1016/j.ssaho.2023.100508

TBS News. (2021, February 1). *Food safety law implementation urged*. Retrieved February 5, 2024, from *The Business Standard*. https://www.tbsnews.net/bangladesh/health/food-safety-law-implementation-urged-195193

The Daily Star (2022, September 7). *OMS rice sale to continue*. Retrieved March 14, 2024, from *The Daily Star*. https://www.thedailystar.net/news/bangladesh/news/oms-rice-sale-continue-3112971

The World Bank (2016, October 9). *Bangladesh: growing the economy through advances in agriculture*. Retrieved March 1, 2024, from The World Bank. https://www.worldbank.org/en/results/2016/10/07/bangladesh-growing-economy-through-advances-in-agriculture

The World Bank (2022, March 15). *Bangladesh: World Bank supports food security and higher incomes of farmers vulnerable to climate change*. Retrieved March 1, 2024, from The World Bank. https://www.worldbank.org/en/news/press-release/2022/03/15/bangladesh-world-bank-supports-food-security-and-higher-incomes-of-farmers-vulnerable-to-climate-change

Timmer, C. P. (2000). The macro dimensions of food security: Economic growth, equitable distribution, and food price stability. *Food Policy*, *25*(3), 283–295.

United Nations. (1987). *Report of the world commission on environment and development: Our common future.*

Wardad, Y. (2020, May 31). *Amphan causes damage to crops worth Tk 6.72b*. Retrieved March 1, 2024, from *The Financial Express*. https://today.thefinancialexpress.com.bd/first-page/amphan-causes-damage-to-crops-worth-tk-672b-1590858231

WFP (2023, February). Bangladesh food security monitoring: Remote household food security survey brief. Retrieved March 13, 2024, from Relief Web. https://reliefweb.int/report/bangladesh/bangladesh-food-security-monitoring-february-2023-remote-household-food-security-survey-brief

Wijerathna-Yapa, A., & Pathirana, R. (2022). Sustainable agro-food systems for addressing climate change and food security. *Agriculture*, *12*(10). https://doi.org/10.3390/agriculture12101554

Index

Printed and bound by CPI Group (UK) Ltd, Croydon, CR0 4YY

15/09/2026

14962486-0002